Applied Wildlife Habitat Management

This book is generously supported by Nancy P. and Thaddeus Edgar "Ted" Paup in recognition of wildlife professionals who are committed to preserving the natural diversity of Texas and in recognition of private landowners who, with the help of professionals, put conservation principles into practice.

Texas A&M AgriLife Research and Extension Service Series

CRAIG NESSLER and DOUGLAS L. STEELE, General Editors

Applied Wildlife Habitat Management

Roel R. Lopez

Israel D. Parker

Michael L. Morrison

Texas A&M University Press   College Station

This paper meets the requirements of ANSI/NISO Z39.48-1992 (Permanence of Paper).
Binding materials have been chosen for durability.
Manufactured in the United States of America

LIBRARY OF CONGRESS CATALOGING-IN-PUBLICATION DATA

Names: Lopez, Roel R., 1969– author. | Parker, Israel D., 1979– author. |
    Morrison, Michael L., author.
Title: Applied wildlife habitat management / Roel R. Lopez, Israel D. Parker,
    and Michael L. Morrison.
Other titles: Texas A&M AgriLife Research and Extension Service series.
Description: First edition. | College Station : Texas A&M University Press,
    [2017] | Series: Texas A&M AgriLife Research and Extension Service series
    | Includes bibliographical references and index.
Identifiers: LCCN 2016039268| ISBN 9781623495022 (hardcover (printed case) :
    alk. paper) | ISBN 9781623495039 (e-book)
Subjects: LCSH: Wildlife management—Textbooks. | Wildlife
    management—Handbooks, manuals, etc. | Habitat (Ecology)—Textbooks. |
    Habitat (Ecology)—Handbooks, manuals, etc. | Wildlife habitat
    improvement—Textbooks. | Wildlife habitat improvement—Handbooks,
    manuals, etc.
Classification: LCC SK355 .L66 2017 | DDC 333.95/4—dc23 LC record available
    at https://lccn.loc.gov/2016039268

General editors for this series are Craig Nessler, director of Texas A&M
AgriLife Research, and Douglas L. Steele, director of the Texas A&M AgriLife
Extension Service.

Contents

Preface

Our primary goal in this book is to provide a practical guide for users with many levels of expertise in the area of wildlife habitat management. We have found that foundational natural resources textbooks are often geared toward the needs of college students. As educators, we find these books enormously useful because they lay a framework of theory and methodology. However, we hoped to accomplish something a bit different in this book. We wanted to create a book that could successfully serve in the classroom but then follow each student into the field as he or she begins careers in habitat management. Rather than sit on a shelf as an occasional reference (and memento of those college years) or sold by the student the moment the semester ends, we hoped to create a book that would be actively referenced. We might then measure success based on how many classrooms and field trucks this book inhabits.

We also wanted to be able to directly hand this book to current on-the-ground managers such as landowners or wildlife biologists who would hopefully find it current, insightful, and useful. This required us to write a book that explains the foundational aspects of wildlife habitat management but maintains a coherent thread and emphasis on application. This translation of theory to practice serves two important roles. First, it ensures that college students receive practical education that is immediately useful on graduation. When we present a method or theory, we attempt to connect it to management scenarios and actions. Second, it could aid in connecting the academic and research worlds with the applied realm. In other words, we thought this book needed to translate emerging research and management ideas into a language that is most useful to on-the-ground managers with years or decades of practical experience.

Ultimately, we will have succeeded in our goal if readers ranging from college undergraduates to on-the-ground practitioners find this book useful. This book is unique in that it provides a step-by-step guide for habitat management that is adaptable to a multitude of environmental settings. It avoids a cover-type or geographic-specific approach by laying out the basic ecological principles applicable to any project and stepping through all of the various sampling designs, measurement techniques, and basic analytical methods needed to develop a project, including writing the actual plan. We build from a review of central ecological concepts, presented in a manner accessible to undergraduates and land/resource managers, and lay out all of the tools necessary to develop plans and implement projects designed to enhance wildlife and wildlife habitat. Thus, while not presenting a cookie-cutter approach, we believe that this book can be used as a practical guide for students, land managers, and landowners in developing and implementing plans for habitat modification and desired outcomes.

Acknowledgments

A book only comes together with amazing support. First, we would like to thank our families for their understanding over the last few years while we took time away from home to work on this project. We also thank Alison Lund for her review of an early draft and extensive index support, and Natalie Thunderbolt for her organizational support early in the draft process. Finally, we thank the book reviewers who provided excellent ideas and constructive criticism.

Applied Wildlife Habitat Management

1 Wildlife-Habitat Relationships

Nature has long fascinated humans, as demonstrated in the beautiful Paleolithic Lascaux cave paintings (nearly 20,000 years old) depicting both humans and wildlife species, leading ancient peoples to attempt to reason out the patterns of the natural world. As naturalism blossomed again in the seventeenth century, scientists began observing and recording the natural world in earnest. This precipitated increasingly common "why" questions that required ever more rigorous scientific methods. Habitat research and management eventually evolved out of the milieu and continues to evolve as the needs of science and managers change.

It should surprise no one that humans have been interested in animals throughout recorded history. After all, animals have long served as sources of food, clothing, transportation, and companionship. In fact, certain animals were also to be strictly avoided because humans became a food source in the natural world! Although hunting wild animals continues into the present day, there has been a long history after domestication of animals to secure a more reliable and controllable food source. Additionally, people were curious about animals, including their natural habits, behaviors, movements, and other life history traits. Aristotle was one of the first *naturalists*, who wrote about the natural lives of animals, recording his observations and speculating on why animals engaged in certain behaviors.

Naturalists recorded little information for over 1,500 years following Aristotle's death. Not until the seventeenth century did naturalists begin to reappear, including such notables as John Ray (seventeenth century) and Carl Linnaeus (eighteenth century). Naturalists of the seventeenth and eighteenth centuries were interested primarily in identifying and cataloging new animal species, which was understandable since no thorough assessment had ever been conducted in recorded human history. After all, it is difficult to study things when you do not even know what to call them.

During the eighteenth century people began making specific trips, or expeditions, into unexplored lands, usually with the primary intent of mapping trade routes and locating needed resources for a growing human population. It was not unusual for an expedition to include a scientist (naturalist) whose job was to collect specimens and record detailed notes on the fauna and flora encountered.

Explorations by naturalists continued into the nineteenth century. Indeed, such explorations continue today, especially in remote and thinly populated portions of the world. This does not mean that we have complete catalogs of animal distributions in the more developed countries. In the United States, for example, large portions of the western Great Basin have received few if any systematic surveys for plants and animals. Furthermore, the distribution and abundance of many animal groups are poorly known even in otherwise well-studied regions: bats, for example.

With the continued natural history explorations came an increasing interest in why animals occurred and behaved as they did. Of course, most prominent in the explosion of interest in natural history was the work by Charles Darwin, an English naturalist. The data collected by Darwin on animal distributions led to his theory of evolution by natural selection, which served as the foundation for the field of animal and plant ecology.

As summarized by Morrison, Marcot, and Mannan (2006), curiosity about how animals interact with their environment led to detailed descriptions of the distribution of animals along environmental gradients (variance of environment through space) or among plant communities (Merriam 1890). During the early 1900s, biologists hypothesized that climatic conditions and availability of food and sites to breed were the primary factors determining the distributions of animals they observed (Grinnell 1917a). Other biologists, however, concluded that the distribution of some animals could not be explained solely on the basis of climate and essential resources. David Lack (1933), for example, proposed that birds recognized certain features of environments that caused animals to occupy a specific location. The writings of scientists such as Grinnell and Lack laid the foundation for the concept of *habitat selection*, a multistep process by which animals innately select certain general features of a broad area (e.g., a type of forest or wetland) and then can use learned experience to make more subtle adjustments of where to feed or breed (Svardson 1949; Hilden 1965). Habitat selection, therefore, is generally seen as involving several levels of discrimination, spatial scales, and a number of interacting factors. The study of habitat selection grew rapidly throughout the twentieth century and continues unabated to date.

Habitat Terminology

Because this is a book on wildlife habitat ecology and management, we must first be clear about our terminology. The nonscientist understands that the term *habitat* refers to a physical location and associated conditions. Various human endeavors designed to enhance the life of other people have adopted this term (e.g., Habitat for Humanity builds homes for people). Unfortunately, the term is somewhat misunderstood and also misused in both ecological and management writings.

A review of even a few papers concerned with the subject shows that the term is used in a variety of ways (Hall, Krausman, and Morrison 1997; Morrison and Hall 2002). Frequently habitat is used to describe an area supporting a particular type of vegetation or, less commonly, aquatic or lithic (rock) substrates. This use probably grew from the term *habitat type*, coined by Daubenmire (1976, 125) to refer to "land units having approximately the same capacity to produce vegetation."

We follow the definition adopted by Morrison, Marcot, and Mannan (2006) in which *habitat* is a concept associated with a particular species, sometimes even with a particular population, of plant or animal. Habitat, then, is an area with a combination of resources (food, cover, water) and environmental conditions (temperature, precipitation, presence or absence of predators and competitors) that promotes occupancy by individuals of a given species or population and allows those individuals to survive and reproduce. Numerous authors have, however, applied modifiers to further describe the term, including "high quality," "marginal," "transitional," "optimal," and "suitable" (Hall, Krausman, and Morrison 1997).

The definition of habitat as species- or population-specific is a core concept of meaningful habitat management. If you cannot specifically define and study habitat, how can you hope to appropriately manage it? Thus, managing vegetation and other environmental features will likely fail to adequately account for the desired assemblage of wildlife. Failing to simultaneously manage for plants and animals is a hit-or-miss strategy; managing vegetation and other environmental features will address habitat for some animal species but will ignore—all or in part—habitat for other species (some of which might not be desired). There are a few key habitat-related terms and concepts needed to understand, quantify, and manage habitat (our terminology follows Morrison, Marcot, and Mannan 2006; and Morrison 2009).

We define *habitat use* as the way an animal uses (or consumes) physical and biological components (resources) within a habitat. *Habitat selection* is the hierarchical process that animals use to make decisions about where to live by utilizing a combination of learned experiences and instinct (D. Johnson 1980; Hutto 1985). *Habitat preference* is restricted to the consequence of the habitat selection process, resulting in the disproportional use of some resources over others.

We also need to know how much habitat occurs in the area of interest. This is critical for managers to properly manage the area. Thus, *habitat availability* refers to the accessibility and ability of an individual to obtain physical and biological components of the environment (area of interest). Habitat availability is not synonymous with the amount or abundance of these resources; *habitat abundance* is restricted to the quantity of resources in the habitat, irrespective of the organisms present (Wiens 1984). It is, of course, difficult to assess resource availability from an animal's perspective. We can, for example, measure the abundance of food for a particular species within a given area, but we cannot know that all of the food in the habitat can be obtained by an animal because there are many factors that restrict its availability. A simple example is plant material such as leaves or fruit that is beyond the reach of an animal; it is thus unavailable as food.

Habitat quality describes a habitat's ability to provide resources for basic survival, reproduction, and individual or population persistence. Habitat can be assigned a rating ranging from low to high to signify its level of appropriate conditions (Morrison 2009). Habitat quality must be linked with demographic features if it is to be a useful measure. Thus, a habitat does not have an inherent quality; rather, quality is assigned by the observer (scientist) after measuring how well animals survive and reproduce during a specific period of time (e.g., breeding season). Van Horne (1983) was the first to popularize the idea that animal density can be a misleading indicator of habitat quality. Thus, although animal abundance can be correlated in some cases with habitat quality, the quality itself should be based on demographics of individuals or populations.

Macrohabitat and microhabitat are frequently used in wildlife ecology to describe the extent (spatial scale) of an area. *Macrohabitat* refers to large spatial scale (landscape scale) with large habitat features such as vegetation associations (D. Johnson 1980; Block and Brennan 1993). *Microhabitat* refers to smaller spatial scales, such as a specific valley, with finer-scaled habitat features, such as a tree within the valley. Depending on the level of complexity desired by the manager, different habitat scales are used to define, study, or even manage an area.

The distribution of animals is also intimately tied to the concept of *niche*. Niche has been defined in multiple ways over time and continues to be the subject of much discussion. As reviewed in Morrison, Marcot, and Mannan (2006), Grinnell (1917b) introduced the term when explaining the distribution of a single species of bird. His assessments included spatial considerations (e.g., reasons for a close association with a vegetation type), dietary dimensions, and constraints placed by the need to avoid predators. In this view, the niche included both positional and functional roles in the community. Elton (1927) later described the niche as the status of an animal in the community and focused on trophic position and diet. Unfortunately, Elton's concept required, in part, that a *community* of animals existed in nature, a community being a group of different species formed by both habitat conditions and behavioral interactions (e.g., competition). Grinnell's concept, however, did not require such a community of animals. Schoener (1989) noted that the concepts of Grinnell and Elton have much in common, including the idea that a niche denotes a specific location in the environment, dietary considerations, and predator-avoiding traits.

The observed population distribution of a species represents its *realized niche*. This term, however, does not include the full range of area under which the species could potentially be found. This observed distribution of a species differs from its potential distribution, the *physiological* or *fundamental niche*, because the species is limited by biotic factors such as competitors or varying levels of resources across a habitat. Being able to quantify or at least understand the physiological niche of a species gives us valuable information for managing that species.

Habitat is a core concept for developing general descriptors of the distribution of animals. However, it can provide only limited insight into factors responsible for animals' interaction with habitat (e.g., habitat selection, resource availability) because habitat alone does not describe the underlying mechanisms that contribute to these interactions. Understanding and utilizing these key terms and concepts will assist managers in better defining the possibilities and constraints of the habitats that they manage.

Spatial and Temporal Aspects of Habitats and Wildlife

Scientists are recognizing that their understandings of wildlife-habitat relationships are scale-dependent, due to the different scales at which all animals (including humans) operate (Wiens 1989). D. Johnson (1980) and Hutto (1985) popularized the concept that animals select habitat through a gradual and hierarchical process: (1) an initial selection decision is made at the geographic range; (2) an animal selects a location to occupy within that larger range (the home range); (3) specific components within its home range are selected (e.g., a particular range of canopy cover); and (4) an animal procures resources within these various microsites. By quantifying this hierarchy of habitat selection from broad to specific, we are able to understand increasingly detailed information about animals. At broader spatial scales we can measure general characteristics of species such as presence or absence, but we are usually unable to quantify survival or reproductive success. Alternatively, information we obtain at the foraging sites informs us about food habitats and behaviors. Thus, we must match the goal of a study with the appropriate scale (fig. 1.1). Table 1.1 lists five dimensions of scale along with guidelines suggested by Morrison, Marcot, and Mannan (2006) for three levels of magnitude of each dimension.

Often, confusion exists concerning the terms *scale* and *level*. For example, the level of biological organization pertains to ecosystems, communities, assemblages, species, individuals, or gene pools, as well as classification levels of vegetation communities such as plant associations, vegetation types, and ecoregions (Bailey 2005). A landscape (larger range) study might pertain to a fine level of biological organization, such as an inventory of ecotypes, but be implemented across a broad geographic scale, such as a drainage basin. In this way, dimensions of scale may be applied at different magnitudes for a given purpose (Morrison, Marcot, and Mannan 2006).

Landscape ecology is the scientific study of species, communities, and ecosystems across geographic areas, typically defined by hydrologic and administrative boundaries (Morrison, Marcot, and Mannan 2006). Although most wildlife species do not strictly adhere to the car-

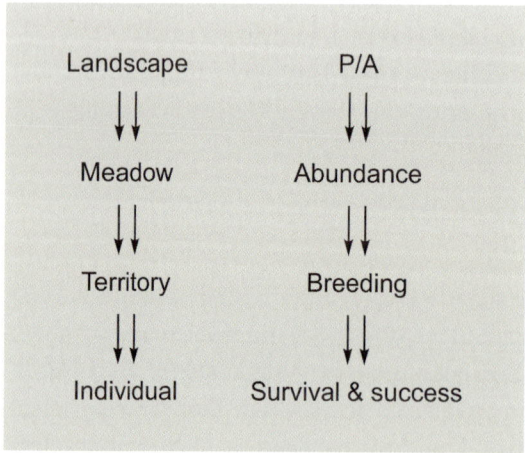

Figure 1.1. Relationship between the measurement of spatial extent and the appropriate measure of animal performance. From Morrison (2009, fig. 4.1).

Table 1.1. Aspects of scale and examples at three levels of magnitude

Aspect of scale	Broad-scale magnitude	Midscale magnitude	Fine-scale magnitude
Geographic extent	Entire major drainage basins or entire ecoregions ("large-scale" study)	Subbasins or local watersheds or more local physiographic provinces; large groups of vegetation patches ("medium-scale" study)	Areas smaller than subbasins or local watersheds; small groups of individual vegetation patches or substrates ("small-scale" study)
Map scale	"Small-scale" maps, e.g., ≥1:1,000,000	"Medium-scale" maps, e.g., 1:100,000	"Large-scale" maps, e.g., 1:24,000
Spatial resolution	Typically coarse-grained, such as for characterization of vegetation and environmental conditions at 1 km^2 pixel size	Environmental patches within subbasins or local watersheds, e.g., ≥ 10 ha	Fine-grained, e.g., <10 ha patches
Time period	Paleoecological past and evolutionary future	Historic past and approximately one century into the future	Very recent management past and current conditions projected only a few years into the future
Biological organization	General abundance of vegetation communities, cover types, and structural stages; mapped locations of ecoregions or ecosystems; inference to broad ecological communities and special assemblages; "coarse- filter" elements	Distribution and abundance of individual species or species groups	Species; gene pools or demes; subspecies or varieties; morphs or ecotypes; "fine-filter" elements

Source: From Morrison, Marcot, and Mannan (2006, table 8.1).

tographer's (mapmaker's) or administrator's boundaries, they often respond to broad landscape patterns as well as to individual patches of resources and environments. A landscape approach to assessment and management of wildlife-habitat relationships is useful for managers because some ecological processes become evident only at a landscape scale. Such landscape processes include biogeography of species distributions, hydrology of surface and subsurface flows, soil erosion, and population dynamics.

Geographic extent is one dimension of scale and typically thought of as "large scale" in landscape, thus a "large geographic or spatial extent." This term has led to much confusion because in cartography it refers to larger values of map-scale ratios—the ratio of a fixed distance on a map (e.g., 2.5 cm, or 1 in) to the distance of actual geographic surface it represents—and thus to maps that generally cover a small geographic extent. Mapping ecological entities varies by map scale. As a result, we follow Morrison, Marcot, and Mannan (2006) and suggest replacing the ambiguous "large scale" with "large geographic extent," and "small scale" with "small geographic extent."

Another dimension of scale is *spatial resolution*, which can range from images with coarse-grained pixels to those with very fine-grained resource patches or point locations of conditions. A small-scale map (e.g., a scale of 1:2,000,000) covering a major drainage basin might be represented either by coarse-grained or finer-grained spatial resolution. It is important to denote the spatial resolution as well as geographic extent and cartographic map scale of a particular map image (Morrison, Marcot, and Mannan 2006).

Time period is another dimension of scale that is not often examined, even though time is critical to interpreting geographic data. This element is especially important to consider when making habitat management decisions because natural and human-induced disturbances are frequent and can be beneficially analyzed using temporal scales. Disturbances should be quantified by their duration and frequency to recognize and analyze patterns. The time period over which data were gathered for geographic analysis in landscape studies is important to report, as some ecosystem processes, including disturbance regimes, may occur beyond the time period studied.

Disturbances and Disturbance Ecology

The field of *disturbance ecology* concerns the dynamics of habitats in landscapes and addresses topics across scales of space and time, including soil dynamics, fire ecology, vegetation succession, meteorology, climatology, and paleoclimatology. As reviewed by Morrison, Marcot, and Mannan (2006), the effects of disturbances on wildlife should be depicted by their frequency, intensity, duration, location, and geographic extent. Understanding these aspects of disturbance allows us to predict how wildlife is likely to respond to both natural and human-induced disturbances. Many animal species, such as nomadic species, likely evolved with native disturbance regimes and are able to take advantage of resources that shift locations through time. However, if human activities change the natural dynamics of disturbances, some animal species may be unable to adapt and could decline in numbers and distribution. For all categories of disturbance, the ability of the land (from very localized to broad spatial extents) to recover—that is, return to predisturbance conditions—depends largely on the severity of the disturbance, the ability of aggressive (and often exotic) species to become established, and other local and regional factors (e.g., prevalence of drought conditions). Four basic categories of disturbances are identified along a range of intensity and geographic extent (table 1.2).

Type I Disturbance

Type I disturbances can bring very different changes to environments and species composition. Hurricanes, for example, can have major effects on wildlife habitat and populations, and in some cases habitat refugia are important protection zones for populations. Volcanoes are another important example of Type I disturbances because eruptions can remove surface vegetation and wildlife, alter soil chemistry and water regimes, impact air quality and temperature, and change surface topography. In fact, volcanic eruptions have proven important in a variety of natural resource management areas, including colonization

Table 1.2. Types of disturbance shown by intensity and geographic area affected

		Geographic area affected	
		Widespread (>1,000 ha)	Local (1–1000 ha)
Intensity of disturbance	High	Type I Major environmental catastrophe (volcanoes, major fires, hurricanes)	Type II Local environmental disturbance (wind, ice storms, insects, disease)
	Low	Type III Chronic or systematic change over wide areas (predators, competition, forestry, regional climate changes)	Type IV Minor environmental change (local fires, developments)

Source: From Morrison, Marcot, and Mannan (2006, fig. 8.5).

(creation of new islands) and recolonization (scarified areas) research and development of the theory of island biogeography (the study of species richness). This is important as wide-scale, human-induced disturbances (such as urbanization causing habitat loss and fragmentation) are more pervasive, and Type I disturbances can be more impactful in these situations.

Type II Disturbance

A common example of a Type II disturbance is locally intense environmental changes from events such as wind storms, ice storms, and local outbreaks of defoliating insects, which in turn cause the formation of gaps in the vegetation cover. Canopy gaps will undergo local succession of plant species and thus become substantial contributors to overall vertical vegetative structure (layers of vegetation) and species composition. Outbreaks of defoliating insects can directly or indirectly cause local and extensive changes in forest structure and composition. For example, in spruce-fir forests in the northeastern United States, the entire bird assemblage showed functional (behavioral) responses to increasing spruce budworm (*Choristoneura fumiferana*) density through increased foraging. However, only two species also showed numerical responses (through increased reproduction). Researchers concluded that the birds were able to reduce the severity (amplitude) of budworm infestations, although the birds could not prevent an outbreak per se (Crawford and Jennings 1989).

Type III Disturbance

The response of wildlife to Type III disturbances includes changes in species abundance and distribution resulting from regional climate changes. As shown by N. Johnson (1994), birds in the southwestern United States have shifted their breeding range northward with gradual warming of the region. Climate fluctuations lasting only a few years can also have substantial impacts, which may last many years after the climate has yet again changed. A popular example of such short-term climatic events is the El Niño warm-water cycles in the eastern Pacific Ocean, which typically impact the breeding chronology and reproductive success of both seabirds and aquatic assemblages (Grant et al. 2000). Type III disturbances can change species assemblages in ways difficult to predict. This is unfortunate, as much of wildlife management is based on accurate prediction of wildlife population demographics and numbers. It also becomes abundantly clear that the common policy of protecting important wildlife habitat in parks or refuges is critically imperiled if habitat areas change or move.

Type IV Disturbance

Although Type IV disturbances are characterized as being relatively minor and localized, they can greatly affect wildlife and habitats. For example, if an individual tree were to fall in the forest, the microclimate at the forest floor could be altered by the canopy that has now opened up, allowing for more sun to shine through and sun-tolerant plants to become established. Additionally, these canopy openings provide areas of nutritious herbaceous growth for a variety of herbivores, such as white-tailed deer (*Odocoileus virginianus*). These nutritional hotspots are often important in an area such as a forest that provides other essential habitat components like cover but might have limited food sources (e.g., little undergrowth). That same opening may also provide other important advantages, such as sunning areas for forest reptiles or hunting grounds for raptors. Ultimately, habitat heterogeneity is critical to the maintenance of a variety of wildlife species.

Habitat Heterogeneity

The heterogeneity (diversity) of resource patches in landscapes has been discussed by different authors in various ways (table 1.3). We define *habitat heterogeneity* as the degree of discontinuity in environmental conditions across a landscape for a particular species. Keep in mind that *habitat* should be viewed as a species-specific term and that a particular environmental condition may constitute a habitat for one species and a limitation for another.

Discontinuities in environmental conditions can occur as relatively sharp breaks in environmental conditions, known as *ecotones*. *Ecoclines*, or broader gradations in conditions over areas of greater geographic extent, can also represent areas with breaks in resources. Discontinuities can occur naturally, such as changes in soil type or edges of water bodies, or anthropogenically, such as edges of plowed grasslands or burned forests.

Fragmentation refers to the degree of heterogeneity of habitats, usually vegetation patches, across a landscape, particularly related to isolation and size of resource patches. Since it refers to habitat, fragmentation is necessarily a species-specific condition. Unfortunately, "habitat fragmentation" is ingrained in the ecological literature and used with abandon to refer to virtually any sort of heterogeneous condition. Also, many authors (e.g., Bogaert, Farina, and Ceulemans 2005) refer to "landscape fragmentation," which is incorrect, because it relates to environments or resources (habitats for specific species) that become fragmented within landscapes, not entire landscapes. Various kinds or degrees of species-specific habitat heterogeneity exist. In an extreme case, resource or vegetation patches can be isolated into islands surrounded by vastly different and, for certain species, unsuitable conditions (fig. 1.2).

Another kind of heterogeneity is temporal fragmentation, sometimes called *ecological continuity*. This refers to the degree to which a particular environment, such as an old forest, occupies a specific area through time. If an old forest ecosystem is interrupted, such as by widespread forest conversion or cutting, and then allowed to regrow, many of the original species closely associated with such environments may nonetheless be lost.

Table 1.3. Components of habitat heterogeneity in landscapes

Component	Description	Example	Source
Corridors	More or less linear or constricted arrays of environments or habitats in a landscape serving to connect larger patches	Movement corridors in undisturbed riparian woodland for cougars (*Felis concolor*) in Southern California mountains; riparian woodland and shelterbelt corridors in North Dakota supporting populations of migratory birds	Haddad et al. 2003; Mabry and Barrett 2002
Permeability	Degree to which an organism can move among patches within a landscape	Effects of microhabitat and microenvironments within clear-cuts on dispersal of red-legged frogs (*Rana aurora*)	Chan-McLeod 2003; Stamps, Buechner, and Krishnan 1987
Edge effect	Incursion of microclimate and vegetation into a patch, typically forested, from a disturbed edge or opening	Clear-cuts causing reduced tree-stocking density, increased growth, and reproduction of dominant trees; higher tree mortality; and incursion of warmer, drier microclimates into adjacent old-growth forests	Chen, Franklin, and Spies 1992, 1995
Edge contrast	Degree of difference in vegetation structure between two adjacent patches	Great contrast in daily average air and soil temperatures; wind velocity, short-wave radiation, and air and soil moisture differ significantly between clear-cuts and old-growth forests	Chen, Franklin, and Spies 1993

Source: From Morrison, Marcot, and Mannan (2006, table 8.2).

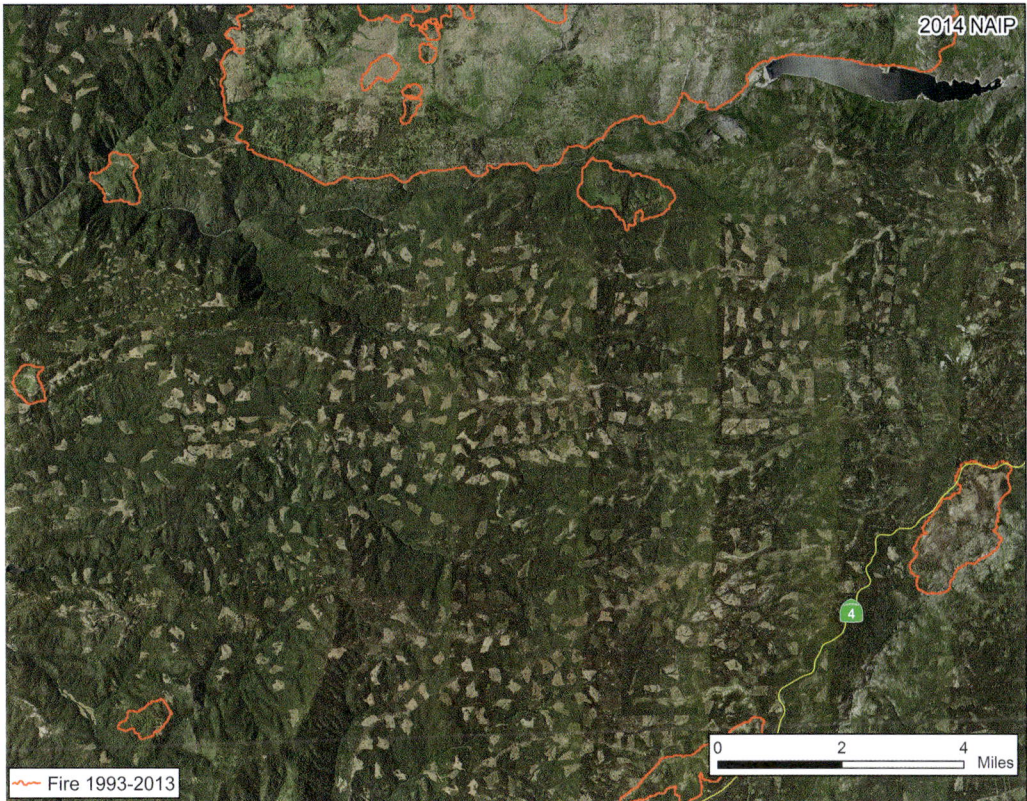

2014 NAIP

0   2   4
Miles

〜 Fire 1993-2013

Figure 1.2. Aerial view showing fragmentation of a landscape in the western Sierra Nevada, California. Fragmentation is caused by the roadways, clear-cuttings, and wildfires shown in the photograph. The scale is 1:100,000. Courtesy of Ross Gerrard, US Forest Service Pacific Southwest Research Station.

Heterogeneity and fragmentation can refer to subtle discontinuities in environmental conditions rather than to changes in gross vegetation structure and successional stage. One example is the horizontal separation of vegetation within a stand, such as among canopies of large trees. This kind of fragmentation has sometimes been called *within-stand patchiness* or *alpha-diversity* of vegetation structure. This kind of fragmentation in forests would likely adversely affect arboreal-dwelling species that require contiguous canopy structures. Another subtle aspect of fragmentation is the vertical separation of vegetation layers such as forest canopies and understories. For example, the degree of heterogeneity of vertical forest stand structure is well known to correlate with bird species diversity.

Landscape ecology in general and disturbance and heterogeneity of habitats in particular have large applications in wildlife habitat management (hereafter management). Morrison, Marcot, and Mannan

(2006) described many "lessons" that wildlife habitat managers (here-after managers) can take away from these areas. First, in most ecosystems, dynamics of vegetation and environmental factors are complex and follow time schedules. Thus, landscapes must be assessed individually to determine the kinds of disturbance types that occur, the local site histories, and the likely vegetation responses to any disturbance regimes caused or altered by management activities. Second, wildlife plays a major role in affecting disturbance regimes and how habitats respond to disturbances, such as through predation or transportation of disturbance agents (e.g., forest pathogens) and herbivory influence on vegetation. Third, management activities can change disturbance regimes and how wildlife may respond behaviorally (functionally) and demographically (numerically) to such changes. Fourth, the future responses of ecosystems that incur disturbances at multiple scales of space, intensity, and time are not very predictable. Fifth, studies of disturbances help guide land and resource management.

Edge

The influence of habitat heterogeneity on animals depends largely on the degree of habitat specialization and the use of intervening environments. One aspect of habitat heterogeneity that has been widely studied is that of edges between vegetation conditions or environmental situations. *Edges* are sharply defined ecotones and broader ecoclines where different communities commingle. Often, plant and animal species richness along edges is greater than within adjacent, relatively homogeneous resource patches.

Historically, managers viewed edges in woodlands and forests as desirable because the greater availability of food enticed game animals such as deer and quail (and probably because the animals were easier to see). Likewise, hedgerows in agricultural fields (strips of shrubs and trees dividing fields) were promoted as a way to increase edge and thus game species (Edminster 1939; Jones, Sieving, and Jacobson 2005). However, species richness is also often relatively high along naturally formed edges, such as riparian corridors (unique plant communities growing along natural bodies of water; fig. 1.3).

The fact that edges have a higher species richness than adjacent areas is likely an artifact of the spatial overlap of species assemblages along edges rather than an ecological preference of organisms for edges. Promoting edges for the sake of increasing species richness will likely have many unintended negative consequences, such as the attraction of both native and exotic predators and brood parasites (e.g., cowbirds).

Figure 1.3. Aerial view of the same area depicted in figure 1.2 but at a scale of 1:4,000. Note the fragmentation caused by the dirt roads and clear-cuttings. There is a clear demarcation between the clear-cuttings and adjacent mature forest, which is a classic example of edge. Courtesy of Ross Gerrard, US Forest Service Pacific Southwest Research Station.

As reviewed by Morrison, Marcot, and Mannan (2006), there are only a few known cases of wildlife species adapted specifically to edge environments, although a large number of species will take advantage of edges as conditions allow. Many cases of so-called edge species may simply be a case of species using different vegetation communities, and their occurrence in edges may be a result of needing proximity to different environments within their home range areas. Examples include Townsend's solitaires (*Myadestes townsendi*), which establish winter territories along piñon–juniper–ponderosa pine edges in the southwestern United States (Salomonson and Balda 1977), and hawks (*Buteo* spp.) coexisting along prairie-parkland ecotones (Schmutz, Schmutz, and Boag 1980).

Researchers have shown that significant edge effects of light, temperature, litter moisture, shrub cover, and other parameters can exist from several meters to a few hundred meters into the adjacent forest or woodland. Chen, Franklin, and Spies (1992) defined a *depth-of-edge influence zone* as the point along a forest edge to forest interior gradient at which a variable returned to a condition representing two-thirds of the interior forest environment. They found that this influence zone ranged 16–137 meters (52–449 ft) from the edge depending on the variable. They also noted that there is no "interior" forest environment in a patch less than 10 hectares (25 ac) if depth-of-edge influence is 137 meters and that edge effects were influenced by topographic position. The concept that small forest patches are "all edge" and have no functional interior environment has caused some managers to discount the ecological value of small forest patches; other studies, however, have shown a high ecological value of retaining remnant patches of native forests within highly managed landscapes (Lindenmayer and Franklin 2002). In a long-term study in Amazon forests, Laurance et al. (2002) showed that edge effects had a central role in the fragmented forest, where even small clearings (<100 m [328 ft] wide) were avoided by many animal species.

Evidence of increased predation around edges has led some researchers to refer to these areas as "predator traps" (Rodewald 2002; Lloyd et al. 2005). Kolbe and Janzen (2002) found that nests of painted turtles (*Chrysemys picta*) that were within about 40 meters (131 ft) of the water's edge had higher predation rates than nests farther away. However, Robinson and Robinson (2001) found that edges created within a forest by selective logging did not have an additive effect. These authors demonstrate how effects of edges and habitat fragmentation on plants and animals can vary substantially by species and location; therefore, creating edge (edge for the sake of increasing species richness) is usually not an effective management goal.

Succession

Ecological succession occurs following a disturbance to an environment. *Primary succession* occurs when a location (regardless of spatial extent) is essentially stripped of vegetation. Examples of primary succession include creation of new land area following volcanic eruptions, glacial activity that can denude valleys to bedrock, and floods. *Secondary succession* refers to the sequential development of plant and animal assemblages following disturbances that leave behind part of the previous flora (plants) and fauna (animals). Examples of secondary succession include secondary forest growth following fire or tree harvesting and vegetation die-off and recovery within drought cycles (fig. 1.4). The specific sequence of recovery of plant and animal species is often difficult to predict because many factors must be taken into account, including the intensity of the impact, the availability of seeds and animals to colonize the site, and the environmental conditions that occur following the disturbance.

Summary

I. Terminology

    A. Terminology in habitat research and management is not always clear, and many of these terms appear in research and management publications with a variety of different (often contradictory) definitions.

    B. It is important that the consumers of those publications have a fundamental understanding of habitat concepts in order to critically evaluate assumptions and conclusions.

    C. The definition of habitat as species- or population-specific is a core concept of meaningful habitat management.

II. Spatial and Temporal Scales

    A. The concepts of space and time are critical components of habitat, and we now understand that management success depends on, among other things, scale.

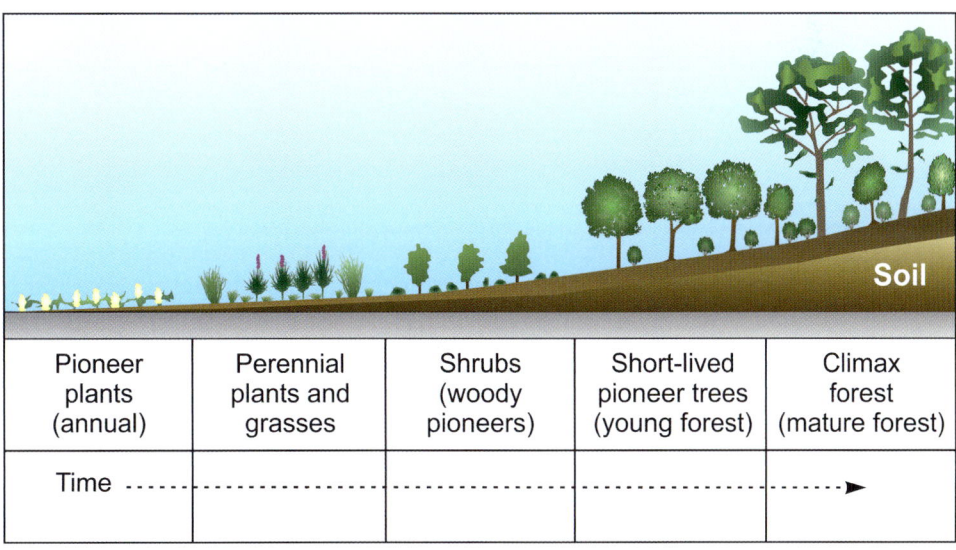

Figure 1.4. Stages of forest succession. Image by M. Morrison.

B. Understanding what influences how animals use habitat at different scales is critical for appropriate management actions.

C. Time influences these decisions as well because of changes in habitat over time (e.g., seasonal changes, lake senescence). Successful researchers and managers must consider the impacts of time on the habitat of concern.

III. Disturbance Ecology

A. From large (Type I) disturbances that can reset succession over large areas to localized (Type IV) disturbances that occur over small areas and short time frames, scale and time are important components of disturbance.

B. This concept is critical to habitat management because wildlife reacts to disturbance in a variety of ways.

C. Management actions often attempt to mimic natural disturbance regimes such as timber harvests, roughly copying natural canopy openings. This is important because disturbance contributes to habitat heterogeneity, which greatly impacts a variety of wildlife species.

D. Discontinuities in environmental conditions pervade the natural world, such as changes in soil types or edges of water bodies, and impact important habitat components, such as edge and succession. These concepts are critical to management actions such as wetland conservation, clear-cut revegetation, and endangered species recovery.

E. Edge was historically viewed as positive because it attracted game animals like deer. Broader multispecies approaches now view edge as a single component in the management paradigm rather than a goal itself.

Literature Cited

Bailey, R. G. 2005. Identifying ecoregion boundaries. *Environmental Management* 34 (1): S14–S26.

Block, W. M., and L. A. Brennan. 1993. The habitat concept in ornithology: Theory and applications. *Current Ornithology* 11:35–91.

Bogaert, J., A. Farina, and R. Ceulemans. 2005. Entropy increase of fragmented habitats: A sign of human impact? *Ecological Indicators* 5 (3): 207–12.

Chen, J., J. F. Franklin, and T. A. Spies. 1992. Vegetation responses to edge environments in old-growth Douglas-fir forests. *Ecological Applications* 2 (4): 387–96.

Crawford, H. S., and D. T. Jennings. 1989. Predation by birds on spruce budworm *Choristoneura fumiferana*: Functional, numerical, and total responses. *Ecology* 70:152–63.

Daubenmire, R. 1976. The use of vegetation in assessing the productivity of forest lands. *Botanical Review* 42:115–43.

Edminster, F. C. 1939. Hedge plantings for erosion control and wildlife management. *Transactions of the North American Wildlife and Natural Resources Conference* 4:534–41.

Elton, C. 1927. *Animal ecology*. London: Sidgwick and Jackson.

Grant, P. R., B. R. Grant, L. F. Keller, and K. Petren. 2000. Effects of El Niño events on Darwin's finch productivity. *Ecology* 81:2442–57.

Grinnell, J. 1917a. Field tests of theories concerning distributional control. *American Naturalist* 51:115–28.

———. 1917b. The niche-relations of the California thrasher. *Auk* 34:427–33.

Hall, L. S., P. R. Krausman, and M. L. Morrison. 1997. The habitat concept and a plea for standard terminology. *Wildlife Society Bulletin* 25:173–82.

Hilden, O. 1965. Habitat selection in birds. *Annales Zoologici Fennici* 2:53–75.

Hutto, R. L. 1985. Habitat selection by nonbreeding, migratory land birds. In *Habitat selection in birds*, edited by M. L. Cody, 455–76. Orlando, FL: Academic Press.

Johnson, D. H. 1980. The comparison of usage and availability measurements for evaluating resource preference. *Ecology* 61:65–71.

Johnson, N. K. 1994. Pioneering and natural expansion of breeding distributions in western North American birds. *Studies in Avian Biology* 15:27–44.

Jones, G. A., K. E. Sieving, and S. K. Jacobson. 2005. Avian diversity and functional insectivory on north-central Florida farmlands. *Conservation Biology* 19 (4): 1234–45.

Kolbe, J. J., and F. J. Janzen. 2002. Spatial and temporal dynamics of turtle nest predation: Edge effects. *Oikos* 99 (3): 538–44.

Lack, D. 1933. Habitat selection in birds with special reference to the effects of afforestation on the Breckland avifauna. *Journal of Animal Ecology* 2:239–62.

Laurance, W. F., T. E. Lovejoy, H. L. Vasconcelos, E. M. Bruna, R. K. Didham, P. C. Stouffer, C. Gascon, R. O. Bierregaard, S. G. Laurance, and E. Sampaio. 2002. Ecosystem decay of Amazonian forest fragments: A 22-year investigation. *Conservation Biology* 16:605–18.

Lindenmayer, D. B., and J. F. Franklin. 2002. *Conserving forest biodiversity: A comprehensive multiscaled approach*. Washington, DC: Island Press.

Lloyd, P., T. E. Martin, R. L. Redmond, U. Langner, and M. M. Hart. 2005. Linking demographic effects of habitat fragmentation across landscapes to continental source-sink dynamics. *Ecological Applications* 15 (5): 1504–14.

Merriam, C. H. 1890. Results of a biological survey of the San Francisco Mountains regions and desert of the Little Colorado River in Arizona. USDA Bureau of Biology, *Survey of American Fauna* 3:1–132.

Morrison, M. L. 2009. *Restoring wildlife: Ecological concepts and practical applications*. Washington, DC: Island Press.

Morrison, M. L., and L. S. Hall. 2002. Standard terminology: Toward a common language to advance ecological understanding and applications. In *Predicting species occurrences: Issues of scale and accuracy*, edited by J. M. Scott, P. J. Heglund, and M. L. Morrison, 43–52. Washington, DC: Island Press.

Morrison, M. L., B. G. Marcot, and R. W. Mannan. 2006. *Wildlife-habitat relationships: Concepts and applications*. 3rd ed. Washington, DC: Island Press.

Robinson, S. K., and W. D. Robinson. 2001. Avian nesting success in a selectively harvested north temperate deciduous forest. *Conservation Biology* 15 (6): 1763–71.

Rodewald, A. D. 2002. Nest predation in forested regions: Landscape and edge effects. *Journal of Wildlife Management* 66 (3): 634–40.

Salomonson, M. G., and R. P. Balda. 1977. Winter territoriality of Townsend's solitaires (*Myadestes townsendi*) in a piñon–juniper–ponderosa pine ecotone. *Condor* 79:148–61.

Schmutz, J. K., S. M. Schmutz, and D. A. Boag. 1980. Coexistence of three species of hawks (*Buteo* spp.) in the prairie-parkland ecotone. *Canadian Journal of Zoology* 58:1075–89.

Schoener, T. W. 1989. The ecological niche. In *Ecological concepts*, edited by J. M. Cherrett, 79–113. Oxford: Blackwell Scientific.

Svardson, G. 1949. Competition and habitat selection in birds. *Oikos* 1:157–74.

Van Horne, B. 1983. Density as a misleading indicator of habitat quality. *Journal of Wildlife Management* 47:893–901.

Wiens, J. A. 1984. The place of long-term studies in ornithology. *Auk* 101:202–3.
———. 1989. *The ecology of bird communities*. Vol. 1, *Foundations and patterns*. Cambridge: Cambridge University Press.

2 Environmental Measurements

Why are wildlife managers interested in sampling vegetation and associated wildlife populations on a given unit of land? Simply put, these three components—vegetation, wildlife populations, and land—are the basic building blocks in wildlife management. First, the measurement of vegetation can assist in determining both the quantity and quality of wildlife habitat and can range from a single plant species to a community of several species. Collectively, vegetation provides many of the components of a habitat—food, cover, and water—that jointly sustain viable wildlife populations.

Second, estimating the associated abundance of wildlife populations is important in evaluating common goals in management. Population size can be considered the "currency" to managers that determines the progress in reaching these management goals. Basic management goals may include increasing population numbers (e.g., rare or endangered species), reducing population numbers for undesirable pest species, or harvesting population numbers at a sustained yield (e.g., game species; Caughley and Sinclair 1994). Thus, estimating the abundance of wildlife populations in conjunction with vegetation measurements is important in making both informed and objective decisions as a manager.

Finally, the collection of vegetation and animal population data and associated estimates is often conducted within some area of interest. Vegetation measurements can be expressed as density (plants per unit area); thus, being able to estimate the area sampled or map the property being managed is as important as sampling vegetation and estimating animal populations. A basic understanding of land measurement is an important skill set for managers and better contextualizes vegetation and animal measurements. Managers can also utilize various tools such as maps or aerial photography, linear measurements (using measuring tapes), and direction measurements (using a compass) to define areas of interest. Understanding these three building blocks of wildlife management will help managers appropriately assess and manage their wildlife and associated habitats. This chapter reviews approaches commonly used in sampling vegetation and animal populations along with basic principles of cartography.

Scales and Terminology

Vegetation, populations, and land are foundational to effective management and, more specifically, the development of management plans. Using sampling techniques allows managers to make inferences about populations (both plants and animals) that can assist in the management process. This chapter presents techniques for measuring, or sampling, common variables related to vegetation populations and their role in land assessment. While there are a number of resources regarding specific practices of these subjects, we review the most commonly used approaches and outline general considerations for managing wildlife populations and their habitats.

It is also important to revisit the idea of spatial scale when discussing vegetation measurements and, to some degree, estimating wildlife populations. Often, a combination of methods is employed; this multisampling approach blends the use of techniques that might be considered broad-scale (e.g., aerial photography) with those that might be considered meso- (intermediate) or micro-scale (e.g., quadrants or line transects). For example, characterizing available habitat for a species might include mapping general cover types (broad scale) and sampling tree and understory structure (meso and micro scales). We also categorize land measurement concepts and approaches under the broad scale. Meso- and micro-scale measurements are discussed in the "Vegetation Measurement" section. Animal measurement is organized by general approaches commonly used in estimating wildlife populations.

First, we present a brief review of terminology important in understanding concepts presented in this chapter. A *population* is a group of plants or animals that occupies an area of interest. *Population abundance* is the number of individual plants or animals (note no reference to area sampled). *Population density* is the number of individuals per unit area, which is more difficult to estimate because the effective area (the area where measurements were obtained) also needs to be estimated. *Relative density* is the ratio of population densities (e.g., density is two times larger in study area A than area B). A *census* is a complete count of an entire population of animals, such as an annual count of whooping cranes. A *population estimate* is a numerical approximation of true population size calculated from a sample. Sampling methods often involve capturing or observing animals or combinations of both processes. *Population closure* refers to the absence of births, deaths, emigration, and immigration over the period of interest (usually the period of sampling), whereas an *open population* has one or more of these factors occurring (population size is changing). A *population index* is a statistic related to population size used to compare populations both temporally and spatially (e.g., annual deer surveys). An *accurate measurement* is one whose value is close to the true value (but the true value of any measurement is rarely known). A *precise measurement* of something is one whose value lies very close to the average of a large number of repeated measurements of that same thing. In contrast to accuracy, precision is a measure of repeatability.

Sample Units and Sampling Design: Experimental Units

Due to limits, such as time and costs, the survey of an entire area of interest is usually not possible. Therefore, an *experimental design* can be employed to select a portion of the area to be sampled (*experimental units*) to represent the entire area of interest. By definition, experimental units are homogeneous and should be representative of the population or treatment to which inference is to be applied. Experimental units may be time periods, units of space, groups of animals, or an individual animal. In simple surveys, where a population estimate is to be obtained from a single entity with no treatments or controls, there is only one experimental unit. In more complex surveys, *sample units* are drawn from the experimental units. For example, within a cage of mice, one mouse is specifically treated with food type A in a dietary experiment. In this scenario, the cage of mice would be the experimental unit, while the mouse being targeted with food type A would be the sample unit.

Likewise, if we are comparing abundance between habitat units (different areas), the differing habitats are the experimental units and each survey would be a sample unit drawn from within each of the habitats. In this type of surveying, sample units are the entity from which measurements are obtained. Sample units are often represented by quadrats, transects, or points in sampling surveys. It is important to note that selection of sample units from within an experimental unit should be done using a probability sampling scheme, or *sampling design*, in which every sample unit has some probability of being selected and by which probability can be accurately determined. Without some type of *randomization rule* there is no way to avoid discrimination or favoritism in sample unit selection, resulting in *bias* (inaccuracy) and unrepresentative estimates of *variance* (precision) in the estimate of abundance.

Several sampling designs exist to accommodate particular survey conditions (Cochran 1977). The most common sampling design is *simple random sampling*, where sample units are selected randomly to ensure that each sample unit has an equal probability of being selected. You proceed by subdividing the experimental unit into sample units, and then you can draw lots, flip a coin, roll dice, or use a random number table to select units to be sampled. During random sampling, sample units may be drawn with replacement (e.g., a sample unit is selected and then placed back into the pool of possible sample units, where it may possibly be drawn again) or without replacement (sample units may be selected only once). Because sampling without replacement is more precise than sampling with replacement, it is more commonly used in wildlife management (Caughley and Sinclair 1994).

Stratified random sampling is employed when there are implicit differences in sample units that need to be accounted for in the analysis. For example, differences in habitat quality may produce localized differences in animal density, resulting in increased variance. To reduce variance, a manager can stratify the area by habitat quality, with sample units selected randomly from each habitat type (e.g., stratifying sample areas into burned and unburned areas).

Systematic sampling (or *systematic-random sampling*) is employed to reduce the amount of effort (time or fuel) necessary to navigate between sample units. Systematic sampling typically uses a random start point and then proceeds in an ordered fashion (e.g., a point grid in which a sample is collected every 200 m [656 ft]) until the entire area to be covered is sampled. It has the advantage of ensuring thorough coverage of the area under investigation but is susceptible to an array of problems (Cochran 1977), the most pernicious of which is the possible coexistence of an unknown periodic variation in the population being sampled (Krebs 1999). The periodic fluctuation could match the frequency of a systematic sampling design, resulting in a biased estimate with unrepresentative precision (and unknown to the user).

Several *non-probabilistic sampling* designs that may be used in error have been described in the literature (Cochran 1977; Krebs 1999), such as *accessibility sampling* (sampling along trails or roads because of ease of access; later called *convenience sampling* [Anderson 2001]), *haphazard sampling* (without a plan), or *judgmental sampling* (selected as "typical" or "representative" on the basis of subjective opinion). Even worse, some sample units may be selected because of the greater opportunity to "see more animals," despite the obvious bias that will result. Regardless of cause or origin, non-probabilistic sampling designs are likely to yield biased estimates with levels of precision that are not representative of the area of inference and should therefore be avoided or applied with caution.

Field Tests

Considerable preparation is required before the data collection team goes into the field to conduct sampling surveys. Regardless of the data collector (academic, independent researcher, agency personnel, private landowner) rigor is required to collect good data and make accurate conclusions. An important first step is making a list of supplies and equipment necessary to complete the task. *Data forms* for recording field data should be developed to facilitate simple mathematical analysis with conventional calculators or to facilitate entry onto a personal computer for detailed and complex analysis. In either situation, a set of instruction codes defining what is represented by each number or letter entry should accompany each field data form. It is important and useful to conduct a small-scale *preliminary field test* of a site before initiating full-scale sampling with the entire team. This field test provides the investigator with an opportunity to evaluate and adjust test equipment, sampling methods, and experimental design and make final estimates of the time required to complete the work. Finally, to maximize field efficiency, personnel collecting data in the field should be properly trained. Field assistants should have a thorough understanding of the safe and proper use of equipment, be familiar with the plants and study area, understand the correct methods for collecting and recording data, and thoroughly understand the rationale of the data collection so that, in the investigator's absence, they can make an intelligent and informed decision when an unforeseen situation arises.

Land Measurement

It is important for managers to have knowledge and understanding of the different elements associated with management. *Land measurement* involves the determination of length, elevation, area, volume, and angles, all of which require a *standard unit*. Two common systems of unit measurement include the English system (nonmetric) and the International System of Units (SI; metric). Despite its disadvantages, the English system persists as the primary basis of measurements in the United States. Being familiar with both systems is important for managers.

Surveying has an alternative meaning in regard to land measurement. In this context, surveying means making field measurements used to determine the lengths and directions of lines on the earth's surface. If a survey covers such a small area that the earth's curvature may be disregarded, it is termed *plane surveying*. For larger regions, where the curvature of the earth must be considered or inaccuracies will occur, *geodetic surveys* are required. Managers are generally concerned with plane surveying—the measurement of distances and angles, the location of boundaries, and the estimation of smaller areas.

Linear Distances

Pacing is perhaps the simplest of all techniques for determining distances in the field. Accurate pacing is an asset to the manager. A *pace* is defined as the average length of two natural steps (a count is made each time the same foot touches the ground). Use of a tape or chain is another approach to measuring linear distances. Two people, a head and a rear chainman, are needed for accurate measurement in order to extend the chain (or other measuring tape) over a specific distance. On level terrain, the chain can be stretched on the ground. On rough terrain, plumb bobs (weights that establish nadir at a particular point) may be used at each end of the chain to aid in accurate measurement of each interval. Chain measurements are slow and subject to a large number of errors. Principal sources of error in chaining are (1) allowing the chain to sag instead of keeping it taut at the moment of measurement, (2) incorrect alignment (not keeping on proper compass bearing), (3) mistakes in counting intervals measured, and (4) reading or recording wrong numbers. Alternatively, an *electronic distance measurement* device (EDM) provides accurate distance readings. An EDM determines the round-trip distance between two points in terms of the number of wavelengths of modulated electromagnetic energy (light or microwave).

Directions

The fundamental values associated with determination of direction are *meridians*, *azimuths*, and *bearings* (fig. 2.1). Meridians are any great circles passing through the earth's north and south geographic poles. True north is the direction of the geographic North Pole along the true meridian. True direction is always referenced to a *true meridian*. *Longitude* reflects how far east or west of the zero meridian a point lies. The *prime meridian* passes through Greenwich, England. *Azimuths* are comparable angles measured clockwise from due north, thus reading from 0° to 360°, whereas *bearings* are horizontal angles that are referenced to one of the quadrants of the compass (NE, SE, SW, or NW).

A *compass*, commonly used in determining a direction, consists of a magnetized needle on a pivot point enclosed in a circular housing that is graduated in degrees. Because the earth acts as a huge magnet, compass needles in the Northern Hemisphere point in the direction of the

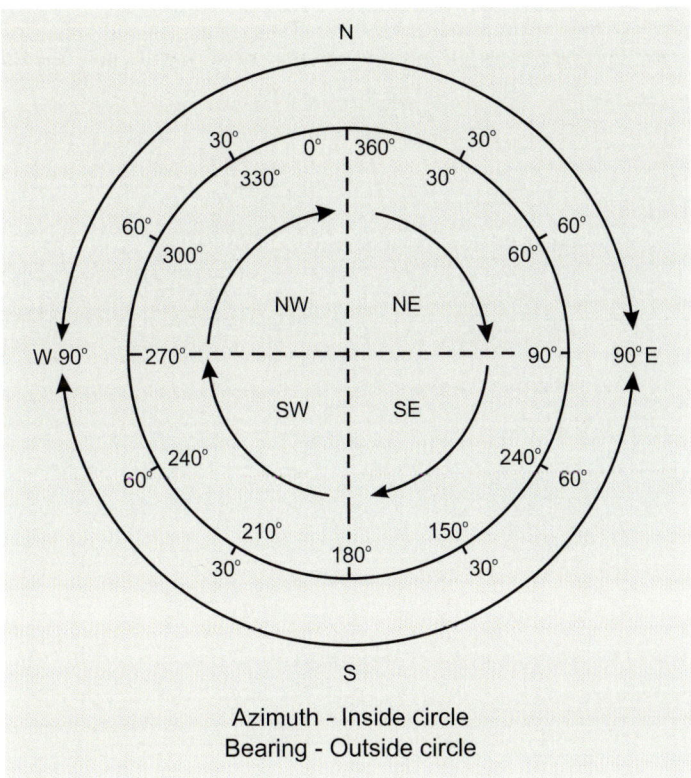

Figure 2.1. Compass rose illustrating two primary ways of expressing direction (azimuth and bearing). Image by Roel R. Lopez.

horizontal component of the magnetic field termed *magnetic north*. If a sighting base is attached to the compass housing, it is then possible to measure the angle between the line of sight and the position of the needle. Such angles are referred to as magnetic bearings or azimuths. In North America, correction may be required for *east* or *west declinations*, the former when the north magnetic pole is east of true north and the latter when it is west of true north (fig. 2.2). Points having equal declination are connected by lines known as *isogons*. The line of zero declination (no corrections required) is an *agonic line*. Declination varies from year to year. In establishing or retracing property lines, personnel collecting data should record angles as true bearings or azimuths. Some compasses allow declination to be set; thus, true readings can be recorded.

Fixed or variable disturbances in the local magnetic field can cause substantial errors in local magnetic directions. Examples include power lines, fencerows, or an automobile. In obtaining a direction, measurement using a compass, often referred to as *frontsights* and *backsights*

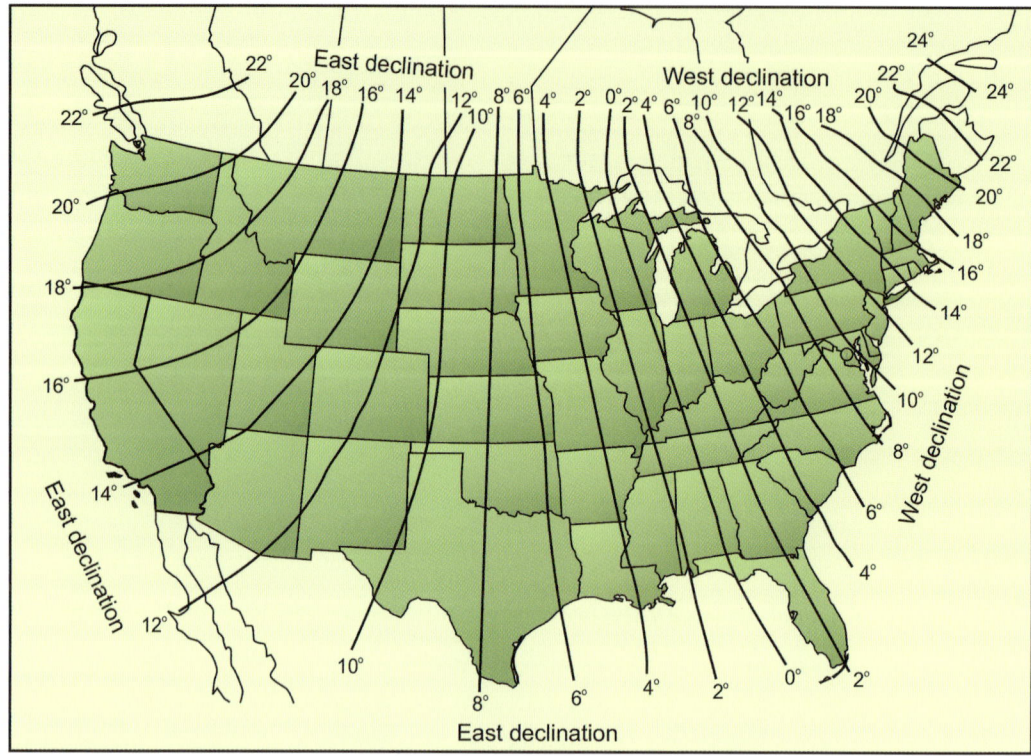

Figure 2.2. Compass declination correction values for North America. Source: NOAA, http://www
.ngdc.noaa.gov/geomag/img/DeclinationMap_US.png.

(180°), should be taken to check the accuracy of the estimate. When
such backsights fail to agree with frontsights, it is likely that some form
of local disturbance is present. It may be necessary to shorten or prolong the traverse line in question so that a new turning point can be
used. To ensure precise compass readings, users should see that the
compass is perfectly level, the sights are properly aligned, the needle
swings freely before settling, and all readings are taken from the north
end of the needle.

Coordinate Systems

Most of us are familiar with the *Cartesian coordinate system* (fig. 2.3).
In this grid system, the horizontal west-to-east grid line is called the
x-coordinate. The vertical south-to-north grid line is the *y*-coordinate.
Because the lines make rectangles, surveyors speak of them as *rectangular coordinates*. If the earth were flat, rectangular coordinates would
serve all map purposes. But a set of coordinates was developed to fit
the globular shape of the earth as closely as possible. There are three

common coordinate systems in land measurement. The system of *latitude* and *longitude* is a circular coordinate system. Any point on the globe is delineated on the north-south axis by a latitude coordinate and on the east-west axis by a longitude coordinate. Both coordinates are measured in degrees, minutes, and seconds. The *Universal Transverse Mercator (UTM) coordinate system* is a worldwide rectangular coordinate system used to locate any point on the earth's surface. UTM coordinates, which are not continuous over the entire globe, consist of two 7-digit numbers, with units in meters. The numbers increase as one moves east and north. The *State Plane Coordinate System* is a rectangular coordinate system. Each state has its own coordinate system with geodetic controls. Coordinates are measured in feet. The selection of coordinate system is related to user needs (e.g., types of maps being used, what kinds of distortions are acceptable).

Area Determination

Various approaches can be used to determine area from the field or a map or photo. In the field, traversing obtains both a direction and linear set of measurements. The most reliable property corner or point of interest is selected as a starting point. The *traverse* may be run clockwise or counterclockwise around the tract from this point. Backsights and frontsights, as well as distances, are taken on each line. Interior angles are computed upon completion of the traverse. If bearings have been properly read and recorded, the sum of all interior angles should equal $(n - 2) *180°$, where n is the number of sides in the traverse. At this point, the tract area can be computed using any of the following methods: *double meridian distance (DMD)*, *double parallel determination*

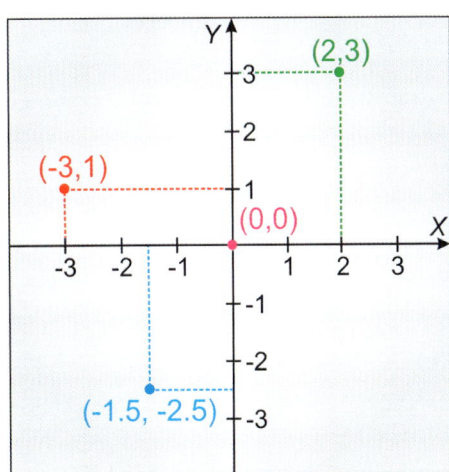

Figure 2.3. A simple Cartesian coordinate system. Image by Joyce VanDeWater.

(DPD), or *rectangular coordinates*. Description of the actual calculations of these approaches can be found in basic land-surveying books.

Often land measurement is not done in the field but from an aerial photograph or map to scale (ratio of a distance on the map to the corresponding distance on the ground) of the property of interest. In this case, several approaches are available to determine the area of interest. In graphical area determination, a grid is overlaid on a photo or map, allowing the amount of actual area per grid to be visually calculated. The total acreage is then determined by counting all small squares within the area's boundary lines. If less than one-half of a square is inside the area boundary, it is ignored. Squares bisected by an exterior line are alternately counted and disregarded.

Similarly, a *dot grid* applies the same basic concepts as a *grid area approach*. If a piece of clear tracing material were placed over a sheet of cross-section paper and pin holes were punched at all grid intersections, the result would be a dot grid. Thus, the dot grid and graphical methods of area determination are based on the same principle: dots representing squares or rectangular areas are merely counted in lieu of the squares themselves.

Another cartographical tool useful in estimating area on a map or photo is the *planimeter*, which has three basic parts: (1) a weighted polar arm of fixed length, (2) a tracer arm hinged on the unweighted end of the polar arm, and (3) a rolling wheel that rests on the map and to which is attached a *vernier scale* (small movable scale). The pointer of the instrument traces the area of interest, and from the vernier scale, the area in square inches is read directly and converted to desired area units.

The *transect method* is a technique for proportioning a known area among various types of land classifications, such as forests, farmland, and urban areas. An engineer's scale (rulerlike device used to measure and transfer ratios) is aligned on a photograph or map so that it crosses topography and drainage at right angles. The length of each type is determined to the nearest 0.25 cm (0.1 in). Proportions are developed by relating the total measure of a given classification to the total linear distance.

Finally, a clever way of estimating area from a map or photo is the *weight-apportioning method*. The areas can be determined by using a copy of a photo or map and directly cutting out and weighing the various units. Assuming that the thickness of the paper is constant, the weight of the unit will be directly proportional to its surface area. This method was developed by the US Soil Conservation Service and has been found to be efficient and accurate.

Scale

In the aforementioned approaches to estimating area using an aerial photo or map, the scale must be known. *Scale* is a ratio of a distance on a photograph or map to its corresponding distance on the ground. The scale of a photograph varies from point to point as a result of displacement caused by tilt and relief. Scale may be expressed as a *ratio*, 1:24,000; a *representative fraction* (RF = photo distance/ground distance), 1/24,000; or an *equivalent*, 0.25 centimeter = 610 meters (1 in = 2,013 ft). The reciprocal of a representative fraction is called the *photo scale reciprocal* (PSR = ground distance/photo distance) and is a commonly used number when calculating distances. Units used in the calculation of RF or PSR must be the same.

Transits and Leveling

In some cases of area determination, there is an interest in taking a measurement in the z-coordinate (third dimension) that reflects changes in elevation. Layout and construction of water lines, roads, and ponds are all dependent on knowing the elevation differences between points. *Transits* or construction *levels* are commonly used in determining changes in elevation. *Leveling* is done to establish the elevations of survey points or the differences in elevation between survey points. Because leveling deals with vertical distances, it is necessary to define what "vertical" is on a round earth. A *vertical line* is defined as a line pointing in the direction of gravity—a plumb line is therefore a vertical line. A *level surface* is everywhere perpendicular to the direction of gravity and therefore curves around the earth. A *horizontal line* is not a line on a level surface; it is a line perpendicular to the direction of gravity. A horizontal line coincides with a level surface at a single point. In small areas, a horizontal line may approximate the level surface. These distinctions are important in different methods of land surveying covered in more detail in other publications.

Remote Sensing

Remote sensing is a set of general land measurement and evaluation tools, which is discussed in our example in relation to biomass. Managers have long sought methods to estimate live herbaceous biomass (biological material derived from living, or recently living organisms) or structure that can be used at finer scales over large areas (Olenicki 2001) and currently rely on either passive or active applications. Ground-based passive sensors or radiometers that measure electromagnetic reflectance from vegetation have been used to measure biomass, amount of green cover, and biochemical constituents (e.g., chlorophyll and nitrogen in a tree canopy) and to classify vegetation in grass, shrub, and forest-dominated systems. Aerial or satellite-based technologies have been used extensively to measure regional vegetation patterns at the largest scale of sampling. Most current land use and habitat coverage are derived from either aerial photography or Landsat thematic mapper multispectral imagery. Aerial photography usually ranges between 1- and 100-meter (3.3- and 330-ft) resolution and requires extensive time and training to utilize correctly. *Landsat* is more easily processed, but the 30-meter (99-ft) resolution image is often too coarse for wildlife management applications. In recent years, managers are finding more applications for active sensors in vegetation measurement that will have a growing role in wildlife management. For instance, radar applications are increasingly used to measure forest canopy and stand characteristics. We also see *LIDAR* (light detection and ranging) or laser altimetry used in canopy measurement for forests and shrubs. LIDAR's capability to characterize three-dimensional canopies at small resolutions holds much promise for wildlife managers.

Vegetation Measurement

Vegetation measurement is important in wildlife habitat management because vegetation provides critical elements of food and cover. Much of wildlife management is actually manipulation of habitat. This often takes the form of vegetation research and management and makes it critical to accurately monitor vegetation metrics. These efforts can prove complicated in the real world. For instance, reduction of invasive, exotic salt cedar (*Tamarix* spp.) is important for water conservation in the Southwest. However, the endangered southwestern willow flycatcher (*Empidonax traillii extimus*) uses salt cedar for breeding in areas where it replaced native willows (*Salix* spp.). Salt cedar removal has the potential to reduce native flycatcher populations and can result in negative consequences if conducted poorly. Prior to any intensive vegetation manipulation, managers must have detailed information about vegetation and wildlife-vegetation relationships. Common metrics or measurements of vegetation can be grouped as frequency, density, cover, and biomass.

Frequency

Frequency is the proportion of sample units in which a species occurs (Bonham 1989). If 25 small plots were examined and a given plant occurred in 10 of those plots, the frequency of that plant would be 40%

(10/25 x 100). Frequency is an easy attribute to estimate because the plant either occurs in the sample unit or it does not, and it is useful in describing distribution of plants within a community. It is also useful for monitoring changes in the plant community over time or comparing different communities (Bonham 1989). If frequency is low (e.g., <10%), plants have an aggregated distribution (occur in clumps) within the community. When the frequency is high (e.g., >80%), plants are uniformly distributed.

Frequency can be obtained via plots, points, and line transects. *Plots* can be square, rectangular, or round and are normally marked or measured with a ruler or tape measure. To measure with *points*, a pin (such as a pointed small-diameter steel rod) is lowered to the ground over herbaceous cover and will either hit or miss a plant. The percentage of hits gives an estimate of the frequency of a species. A single pin (Owensby 1973) or a frame containing several (usually 10) pins (Bonham 1989) can be used to measure frequency. Sample size is a consideration when frequency is estimated. Greig-Smith (1964) recommended that no fewer than 100 sample units be obtained to provide reliable estimates. *Line transects* can vary in length, which is typically determined by the item being measured (discussed later in the chapter). Larger, well-dispersed targets (overstory trees) would be longer than smaller and more localized targets (bunchgrasses).

Density

Density is the total number of objects (individual plants, seeds) per unit area. One advantage of a density estimate is that count data are straightforward to interpret, and the results are comparable to those obtained using other methods (Gysel and Lyon 1980). A disadvantage of a density measurement is that the data are commonly difficult to obtain. Such variability requires larger sample sizes for statistical reliability. Density is a useful and often important measurement for the evaluation of a wildlife habitat but is not in itself an adequate descriptor of a plant community because it does not provide information about how plants are distributed within the community. Approaches to measuring density can include either quadrats, which are defined by fixed boundaries, or plotless methods that have no boundaries.

Quadrats are made of materials with fixed boundaries (e.g., a portable square or rectangle made from PVC pipes), whereas plotless methods are estimated visually. Quadrats require consideration of three characteristics in obtaining a measurement (Bonham 1989): (1) distribution of the plants, (2) size and shape of the quadrat, and (3) number of observations needed to obtain adequate estimates of frequency and density. *Plotless methods* are based on the premise that density can be estimated from the mean area occupied per tree: density (trees/m^2) = 1/mean area (m^2/tree).

The challenge is to estimate the area occupied per tree from distance measurements obtained in the field, including the *closest-individual method*, *nearest-neighbor method*, *random-pairs method*, and *point-quarter method* (Bonham 1989). Of these, the point-quarter method has been widely used in many vegetation types throughout North America. One randomly or systematically selects points within a plant community and measures the distance to the nearest plant within each of four quadrants around the point. Mean area is calculated by squaring the mean distance between points and individual stems (density = 1/d^2). This method can be used to calculate density of all species collectively or density of individual species. The point-quarter method has been criticized because it provides reliable estimates of density only when plants are distributed randomly but not when plants are clumped or uniformly distributed (Oldemeyer and Regelin 1980). The choice of using a plotless method over a quadrat method will depend on the objectives of the study. If the density of one to two species is required, plotless methods appear to be faster than quadrat methods. If the density of all species in the community is desired, the quadrat method is recommended (Bonham 1989).

Cover

Cover is the vertical projection of the crown or stem of a plant onto the ground surface. Managers can use cover to measure the amount of forage for livestock or to describe basal area of merchantable timber. Cover also can be an estimator of biomass when height structure of a community is known. Managers generally look at two types of cover: canopy or crown cover and basal cover. Canopy cover varies within a season or among years. Basal cover, however, is a reliable measurement and is frequently measured at a height of about 2 centimeters (about 0.8 in) on bunchgrasses (Bonham 1989) and at 1.5 meters (5 ft) aboveground for single-stemmed trees (Mueller-Dombois and Ellenberg 1974). Both canopy and basal cover are measured in a variety of ways.

Ocular estimates can be obtained with relative ease in grasslands because of their low profile and height. Ocular estimates are subject to personal bias, however, requiring consistent training and calibration between users. The *line-intercept method* is particularly suited for measuring basal area of bunchgrasses or canopy cover of shrubs, especially in arid or semiarid lands. A line or tape measure is placed between two stakes, and basal width or canopy width of all plants touching the line or tape is measured. Cover is expressed as a percentage of the total length of tape intercepted by vertical projections of the canopy.

The *point-intercept method* measures basal and canopy cover as the percentage of points whose vertical or angled projections intercept vegetation. This method is best suited for estimating cover of herbaceous and low-shrub vegetation. A single pin is lowered toward the ground. The first strike of any part of the vegetation canopy becomes a canopy cover hit; if it strikes the basal area of a plant, it is a basal hit. Percentage of canopy or basal cover is calculated as the total number of hits divided by the total number of pin placements times 100. Many variations of this technique are used, including transects where an observer has a single pin, for example, on his shoe and "hits" are recorded with each step.

The *Bitterlich variable radius method* is a modified point-sampling method developed for use in forestry to measure basal area of trees (Bitterlich 1948). Shrubs or trees are viewed with a clear glass prism that refracts light rays. The observer holds the prism over the sample point while viewing tree stems through the glass and over the top of the prism. The tree trunks appear displaced to one side due to refraction of light passing through the glass. Basal area is calculated by recording the number of trees whose trunks, when viewed through the prism, appear displaced within the trunk line of the actual tree or are still "touching" the main trunk line. The tree is not recorded if the trunk viewed is completely displaced. The stem count per sample point multiplied by the basal area factor (stem count to basal area conversion factor) gives the total basal area of stems (m^2 or ft^2) per unit of area (ha or ac). Basal area factor is a number generally calibrated with a prism (piece of glass).

At times, *tree canopy cover* is an adequate, perhaps even preferred, measure of overstory structure and composition. Many data collectors prefer to use a *spherical densiometer* for estimating tree canopy cover. The spherical densiometer uses a curved, gridded mirror that reflects the overstory at a point and provides estimates of relative amounts of the area covered. The observer levels the densiometer at about chest height and counts the proportion of quarter cells (etched in the mirror) obscured by the reflected vegetation.

Biomass

One of the best indicators of species importance within a plant community is composition based on dry weight (Daubenmire 1968). Managers frequently require data on biomass or standing crop rather than density or cover because biomass is closely related to forage availability and habitat carrying capacity (the maximum, equilibrium number of organisms of a particular species that can be supported indefinitely in a given environment; Bonham 1989). Herein we use the term *biomass* synonymously with *standing crop* to include both live and dead vegetation. Woody biomass and size structure are required to estimate fuel loading, a necessity for formulating fire prescriptions and predicting fire behavior in wildlands. Managers often are interested in measuring biomass of edible components of browse, such as current annual growth, foliage, or twigs. Measurement of biomass can be carried out by three techniques: (1) clipping, (2) ocular estimation, or (3) dimension analysis.

Clipping is a common technique involving the removal of vegetation for analysis. Plant biomass can be measured directly by removing all of the vegetation in a sample plot to ground level and measuring its mass immediately (wet mass) or after air- or oven-drying the sample (dry mass). Mean biomass per unit area can then be estimated as the product of mean biomass per plant (g/plant) and mean density of plants (plants/ m²). This "clip-and-weigh" method can also be used to estimate twig biomass within plots. Clipping all twigs within plots is a highly accurate yet laborious means of measuring browse biomass (Shafer 1963).

Herbage biomass can also be determined by *ocular estimation*. Weighing the plants helps the observer develop more accurate ocular estimates by clarifying the relationship between appearance and biomass. *Dimension analysis* is used in wildlife management for estimating shrub biomass. The technique assumes that plant attributes are related and that one attribute (in this case biomass) can be predicted from another that is more easily measured, such as basal stem diameter, maximum plant height, or various crown dimensions (Whittaker 1965).

Other Attributes

Managers may find that understanding habitat requires evaluation of a variety of other vegetation factors, such as visual obstruction, herbage height, tree dimensions, tree age, and wildlife use of plants. Visual obstruction caused by vegetation may be functionally important to wildlife both as hiding and thermal cover. Managers and researchers have extensively used the measurement of horizontal cover of vegetation in assessing wildlife habitat suitability, habitat preference, impacts of land use practices on wildlife habitats, visibility bias in wildlife surveys, and standing biomass. A variety of different-sized devices have been used to measure horizontal visual obstruction caused by vegetation, including *density boards* (Wight 1939) and *vegetation profile boards* (Nudds 1977). The Nudds board allowed assessment of visual obstruction in vertical units. Horizontal cover is assessed in each interval by viewing the board from 15 meters (50 ft) in a randomly chosen direction. Robel et al. (1970) used a pole-shaped cover board (3 x 150 cm [1.2 x 60 in]) that could be read from a standard distance (4 m [13 ft]) and height (1 m [3.3 ft]) in any direction. The pole was marked in decimeters, and the height of total visual obstruction was recorded. Although ocular estimation techniques are quick to implement, studies that rely solely on them may forfeit accuracy to save on labor and sampling costs.

Vegetation height is probably the easiest attribute of vegetation to measure in grasslands and can often be estimated with high precision. It refers to the tallest portion of a plant or the effective cover height (generally the upper limit of vegetation leafiness). Vegetation height in grassland habitats has an important role in predator deterrence and prey security. Plant height correlates well with other structural attributes of vegetation important to the management of grasslands (e.g., vegetation height relates to foliage density; Higgins and Barker 1982).

Tree dimensions (size and proportions of individual trees) affect the physiognomy (vegetation structure and form) of forested wildlife habitats. A variety of tree sizes, age classes, and structures contribute to habitat complexity and overall diversity of wildlife species. Three common, interrelated measures of tree size are height, crown volume, and trunk diameter. Height of tall trees can be measured using a trigonometric function of horizontal distance of an observer to the trunk of a tree and the angle measured between the horizontal distance and a sighting to the treetop. Crown volume can be determined from similar measurements of minimum and maximum canopy height and canopy diameters measured horizontally. Trunk diameter and cross-sectional area are the most common measurements of tree size because of ease of measurement and high correlation with height and volume. Diameter at chest height (generally 1.4 m aboveground [4.6 ft]) can be measured with a diameter tape placed directly around the circumference of a tree trunk or with calipers. Such data often are summarized as numbers of individuals of species per size class per unit of land area.

Tree age data are beneficial in forest history and dynamics, including predictions of future status (e.g., relationship of age to tree characteristics important for wildlife). Age classification of trees is possible because trunk lateral growth occurs in annular increments related to the seasonality of temperate zone climates (Raven, Evert, and Eichhorn 1986). The increments are especially evident in so-called ring-porous tree species such as oaks (*Quercus* spp.), ashes (*Fraxinus* spp.), and elms (*Ulmus* spp.). Large-pored vascular tissue is formed early in the growing season, followed by small-pored tissue and then termination of growth that year. "Diffuse-porous" angiosperm species like maples (*Acer* spp.), aspens (*Populus* spp.), and birches (*Betula* spp.), have fewer apparent growth rings. Conifers, unlike angiosperms (flowering, seed-bearing vascular plants), have a somewhat different anatomical structure, yet they too typically have easily recognized growth rings. Growth rings can be seen on trunk or stump cross sections.

Vegetation sampling in concert with timber harvest or removal of damaged or dead trees is an easy way to collect such data. Where destructive sampling is not in order, small cylindrical cores can be collected with a wood increment borer. Plant use by herbivores must be quantified due to potential impacts on ecosystem health (e.g., overgrazing can depress desirable plant species). As done for biomass, plant use can be estimated with ocular methods. These require investigator training with ungrazed plants that are clipped to simulate different intensities of grazing. Such estimates vary by individual investigator and may be inconsistent from year to year. Commonly accepted methods of measuring vary from simply counting used or unused stems, to obtaining "before and after" measures of stem lengths, to using regression methods.

Techniques for Sampling Fruits

Fruit abundance data can be quite important when managing for certain species of wildlife that are dependent on annual fruit production. In studies of wildlife food habits, fruits generally are referred to as *mast* and are divided into two categories, "soft" (fleshy exteriors such as berries) and "hard" (dry or hard exteriors such as nuts). Consequently, mast can be defined as the fruits and seeds of all plants, both woody and herbaceous, used as food by animals. Mast is popular with wildlife because it is high in food energy, especially carbohydrates and fats (Goodrum, Reid, and Boyd 1971). Fruit production is impacted by a variety of factors, including plant age, size and genetics, climate, soil, resource competition, previous animal use, and seasonal differences (Schupp 1990). Therefore, management practices that provide for the greatest variety of food-producing plants will assure favorable conditions for the greatest variety of wildlife. This is even more complex because managers must mold strategies depending on mast size, such as large or heavy fruit that often come from bigger trees or smaller fruit from small trees, shrubs, or herbaceous plants.

Large-fruit sampling design depends on whether total mast production or an index of annual mast abundance is desired. Although counts of mast on the forest floor are generally unreliable estimators of mast production due to undocumented wildlife offtake (e.g., eating, storing), such counts, when used with seed traps, can be good estimators of wildlife use of fallen mast. Total counts of mast on trees may be quite accurate for small trees but difficult and time consuming for large trees. Consequently, many researchers and managers have opted to use relatively rapid indices of mast production. Indices based on visual counts (rated incrementally from "none" to "large crop") have the advantage of being quick, permitting rapid assessment of acorn production, but suffer from observer bias. Many kinds of traps have been used to measure mast fall. Polyethylene film traps, square wire-cage traps, and paperboard seed traps were found to be the most efficient, durable, and economical (Thompson and McGinnes 1963).

Small-tree, shrub, and herbaceous fruits are important wildlife foods used by many small rodents, tree and ground squirrels, and game and nongame birds. Abundance of small or *light mast* (e.g., pine seeds) varies from year to year as it does in all fruiting species. The two principal techniques of sampling small or light seed production of trees are placing seed traps in a stand or counting the number of ripening cones on a tree through the use of binoculars. Shrub mast (soft and hard) often is within reach of a biologist and can be counted or harvested directly from the shrub. Like most total enumeration methods, estimating total production of berries can be quite time consuming. Consequently, soft mast production may be characterized for extensive surveys on a scale of relative abundance ranging from 0 to 4.

Managers can also calibrate relative abundance indices to actual production by measuring fruit production on sample plots on which relative abundance is measured. Biologists must remain aware of variation in relative abundance estimates due to differences in observers, regions, or years. Herbaceous mast sampling has not been as well developed as for trees because more plant species are involved, and wildlife that uses those seeds generally is less obvious (e.g., rodents, invertebrates). Samples can be taken from the ground, from traps, or directly from the plant, but these variations are smaller than those used for trees and include fine-screen wire seed traps and traps with integrated adhesives on various substrates such as paper or dishes.

Animal Measurement

Knowledge of the size or density of an animal population is often necessary to manage it effectively. Animal abundance can be measured in two ways: (1) *absolute density*, the number of animals per unit area; and (2) *relative density*, the density of one population relative to another. Although *census* is the total count of animals in an area, here we use the word as an estimate of population size or density (Pierce, Lopez, and Silvy 2012). Closely related to the census is an *index*, a number that is not itself an estimate of population size or density but has a proportional relationship to it, such as deer spotlight counts. In general, population estimates can be divided into three groups: (1) total counts, (2) indices, and (3) population estimates.

Total Counts

Total counts are the complete enumeration of animals in a given area of interest. The idea of counting every animal in a given area has an attractive simplicity to it. Despite known limitations (e.g., suspect accuracy, expensive), total counts do have a place. Total counts on small areas can be derived from intensive surveys, a known number of marked individuals, or other creative means. Examples of total counts include bird dog census (quail, pheasant, waterfowl nests), counts of social groups (flocks of turkeys), drive counts (deer, rabbits, quail), aerial (visual/photographic) census (elk, pronghorn, waterfowl), saturation trapping of sample areas (small mammals), and cleanup or total kill (deer; Pierce, Lopez, and Silvy 2012). Total counts may be possible when counting the number of cows in a small paddock or the number of horses in a small pasture. But for wild populations, it is seldom possible for wildlife managers to obtain a total count of animals.

A drive count is a common strategy in obtaining a total count. As the name implies, counters drive animals to census the total number of animals in a defined area. The technique is best suited for large, easy-to-detect species that inhabit relatively open habitats, such as deer and pheasants. Drivers, spaced along a line, sweep across an area with well-defined boundaries. Additional observers may be placed along the boundaries to count animals that move in or out of the survey area. The census is the sum of the number of animals moving past the line of drivers. McCullough (1979) compared drive counts with population estimates and reported that at low densities, drive counts underestimated the true population, and at high densities, they overestimated the true population. Errors could be as large as 20%–30%. Thus, drive counts are probably best viewed as an index of population size.

Remote Sensing

Thermal infrared (IR) scanners that detect infrared energy emitted by homeotherms (organisms that regulate internal body temperature largely independent of environmental temperature) and vegetation have been proposed as a technique to census animal populations. As early as the 1970s, Parker and Driscoll (1972) suggested that thermal scanning for wildlife was feasible, but appropriate equipment needed to be developed. As the technology developed, a variety of applications have arisen, including aerial surveys of terrestrial (animals that live predominantly or entirely on land) and avian (bird) species and ground-based surveys using forward-looking infrared (FLIR) scanners (Haramis et al. 1985; Bajzak and Piatt 1990; Wiggers and Beckerman 1993; Gill, Thomas, and Stocker 1997; Bontaites, Gustafson, and Makin 2000).

Spot Mapping (Territorial Mapping) and Total Mapping

Spot mapping involves plotting locations of individual birds that are seen or heard on a gridded map during repeated visits to a study area. The technique is most suited to survey birds that regularly sing or call within exclusive territories. Floaters (nonterritorial birds) and young of the year (less than a year old) are usually not surveyed by this technique. The combined data reveal clusters of locations, assumed to represent centers of activity for individual bird territories during the breeding season. The total number of clusters in the study area equals the number of clusters completely inside the area plus the sum of parts of clusters on the boundaries. Total number of birds is estimated by multiplying the number of clusters by mean number of birds per cluster, which is usually two, assuming birds breed in pairs. Total mapping is similar to spot mapping except breeding birds are first trapped and color-banded, prior to surveys to delineate territories. This facilitates the identification of individuals. Total mapping can be extremely accurate but might estimate only conspicuous birds (Pierce, Lopez, and Silvy 2012).

Indices

For animals that are elusive or otherwise difficult to census, methods were developed to census animal indices. *Indices* are typically based on cues or other by-products of animal activity (fecal pellets, nests, burrows, tracks, scrapes, etc.) that are believed to be proportional to animal abundance or density. Animal ecologists have long dealt with statistics that do not actually estimate but are believed to be correlated with animal abundance. Most uses of indices of either abundance or density involve comparisons between populations from the same location at different times or between populations from different locations at the same time. Thus, indices are used to indicate relative differences in abundance. Building on this, an index of density is some attribute that changes in a predictable manner with changes in density (pellet counts to estimate deer density). Indices of density are useful for comparing the density of two populations or for tracking changes in the density of one population from year to year. Without knowing anything about the proportional relationship between the index and population size, we could confidently say that if the index halved or doubled, it would reflect roughly a halving or doubling in animal density.

A density index can be defined as any measure that correlates with density (Caughley 1977). Indices are used most often because of perceived savings in cost, time, and/or labor. Indices differ from population estimation methods in that only relative abundance or relative density can be derived from the indices. Data are usually presented as deer/kilometer, rabbit pellets/meter, or birds/point. Indices can be used to compare animal numbers between treatment and control areas (e.g., disked areas [where soil is disturbed by metal-disked equipment] with non-disked areas) or to compare the same area over time, with the assumption that nothing changes except the relative abundance of the animal being studied.

The probability of catching, counting, or otherwise detecting an animal in sample units from two areas or time periods being compared should be similar. Indices should be standardized for season, time of day, weather conditions, habitat, and observer experience. For multi-species surveys, detection probability will vary with species. Factors such as group status, reproductive cycle, sex and age ratios, and population density will also affect detection probability. For aquatic species, water level, water temperature, and moonlight also affect detection probability.

Consistency and rigid standardization of techniques are crucial when estimating an index. A good observer is not one who gets a high tally but one who has a consistent level of concentration and produces results of high repeatability. Indices or estimates are obtained by sampling, the technique of drawing a subset of sampling units from the complete set and then making deductions about the whole from the part (e.g., counts of sounds produced by males like doves or elk, animal sign such as deer pellets).

Most indices collect frequency (number of individual animals or animal sign) information along transects, at quadrats, or at points. Examples of what type of data can be gathered by index methods include the number of animals seen per kilometer of road, the number of animals present per night at a waterhole, fecal pellets per quadrat, and nest or burrow counts per kilometer of transect. Similarly, a frequency-of-occurrence index collects only presence or absence data. A frequency-of-occurrence index is based on the proportion of sample units, such as scent stations, that contain at least one animal or animal sign.

Most index surveys can be readily modified to provide information needed to use in a population estimation procedure. For example, live catch per unit effort (an index) could be easily modified for a mark-recapture population estimation procedure, or deer seen per kilometer could be easily modified for a line-transect population estimation procedure. Regardless, given the advances in sampling methodology, there are relatively few circumstances where index could not be adequately replaced by a more quantitative method.

Species Distribution

Previous attempts to manage animal populations using indices and complete counts (census) revealed the analytical and practical limitations of these methods. As the size of the area to be surveyed increased, practical limits in available sampling capability were reached, forcing investigators to derive methods for obtaining estimates from samples. Similarly, development of methods for obtaining estimates from samples revealed the importance of sample distribution in relation to species distribution. Resources, and therefore wildlife, are not randomly distributed and thus can create bias in estimates of animal abundance. Problems arise when animal distributions are clumped or when the distribution of samples correlates with the underlying distribution of animals to be sampled. Appropriate survey design is almost always the key component in alleviating this problem, with random sampling or stratified random sampling being the most common remedy. Regardless, we warn that it is prudent to use probabilistic sampling as the easiest alternative for avoiding unforeseeable problems in obtaining estimates.

Population Estimation

Commonly, *known-to-be-alive* or *minimum-number-live* estimates are used to conduct animal population estimates on smaller areas of interest. This type of abundance measurement is conducted by recording initial species sightings and, when possible, resightings of the same animal. The initial count is used to estimate the species' abundance in the area. These estimates tend to underestimate population density; however, an overestimate of density may lead to inappropriate management action, whereas an underestimate may produce inefficient but safe management strategies.

Florida Key Deer

(*Odocoileus virginianus clavium*)

Description: The Lower Florida Keys contain a unique and fragile vegetation structure highly impacted by sea level. Vegetation communities such as mangroves, salt marshes, grasslands, and upland forests are often separated by 1–2 meters (3.3–6.6 ft) of elevation above sea level. These communities support highly diverse flora and fauna. Even small changes in sea level cause dramatic changes in terrestrial vegetation and related wildlife species. The only consistent sources of fresh water are located in the highest elevations. The Florida Key deer is an endangered subspecies of white-tailed deer endemic to the Lower Florida Keys. It is dependent on the upland deciduous and pine forests for cover, fresh water, and forage. Unfortunately, global warming has already begun to raise sea levels, which dramatically impacts vegetation in the Lower Florida Keys. As sea level rises, the amount of upland vegetation declines and decreases Florida Key deer

Florida Key deer. Photo by Roel Lopez.

habitat. Additionally, the only available source of fresh water in the upland areas gradually becomes more saline. The challenge is to predict future Florida Key deer habitat in the face of sea-level rise and determine practical methods of conserving as much habitat as possible.

Scenario: There are between 400 and 600 Florida Key deer on Big Pine Key. This island (~2,500 ha, or ~6,178 ac) serves as the core and largest remaining chunk of habitat. Based on what you have learned from information presented in this book about geospatial analyses, population estimation, and forestry practices, do the following:

1. Develop a comprehensive plan for determining current Florida Key deer habitat. How would you delineate upland forest habitats? How would you confirm which areas Florida Key deer are using? How would you determine the current population density? Be specific and detailed. For example, simply stating that you will use mark-recapture to determine population density is not prescriptive enough. How many individuals would you need to capture? How would you track them or recapture them? What are the drawbacks to this technique?

2. Determine the future habitat in various sea-level-rise scenarios. This requires up-to-date aerials, orthographic maps, vegetation maps, and predicted sea-level-rise maps and expected impacts. What other data would you need? Conduct an online search for the relevant data. You will be surprised at what data are readily available on sites maintained by agencies and municipalities such as the US Geological Survey. Keep in mind these data are collected by various agencies. Expand your search to peer-reviewed journals to find specific data sets related to Florida Key deer. What data were you able to find? Which data were free, and which were protected behind a pay wall?

Population Density

Various methods are available to measure population density, including mark-recapture techniques, quadrat counts, distance methods, sampling with probability proportional to size, and exploited population techniques.

Mark-Recapture Techniques

One way to estimate the size of animal populations is to use a *mark-recapture technique*. As the name implies, individual animals from a population are captured and marked, and then the area is "resampled" to see what fraction of individuals caught carry the marks from the initial capture. The strength of mark-recapture techniques is that they can provide information on birth, death, and movement rates in addition to information on absolute abundance. The weakness of these techniques is that they require considerable time and effort to obtain the required data, and to be accurate, they require a set of very restrictive assumptions (e.g., population closure for Lincoln-Petersen estimator) about the properties of the population being studied. Mark-recapture techniques may be used for open or closed populations. An *open population* is the more usual case, a population that changes in size and composition from births, deaths, and movements. A *closed population* is one that does not change in size during the study period (the effects of births, deaths, and movements are negligible). Following are examples of estimators that exclusively examine open or closed populations.

Lincoln-Petersen Estimator

The *Lincoln-Petersen estimator* is a mark-recapture analytical technique used when the population being studied is assumed to be closed. Commonly this method requires only two visits to the site for sampling if it is conducted during a short period of time to avoid population births, deaths, or animals moving into or out of the surveyed area (population closed). A known number of animals in a study are marked during a short time period, and within a few days a random sample is taken to determine the number marked in the sample. A rule of thumb is to use a different method to obtain the sample than was used to mark the animals, as similar capture methods may bias data because of changes in animal behavior.

The important requirement to mark-recapture methods is that the proportion of marked to non-marked in a sample is the same as in the population. If animals lose their marks, then fewer marked animals would be seen than expected and would cause an overestimation of population abundance. However, factors such as "trap-happy" animals that cause one to see more marked then expected will result in an underestimation of population abundance.

Additionally, investigators must obtain a random sample across the study area that will provide a true ratio of marked to non-marked in the population. For large animals such as deer, which can more easily be trapped and marked along roads, a resight (visual "recapture" rather than physical recapture) survey using randomly placed infrared motion-sensitive cameras is ideal, especially if neck collars are used to mark the deer. Also, the Lincoln-Petersen estimator does not require that animals be individually marked, making this ideal when photos may not capture a good angle of the marked animal. However, additional information (movements, survival) can be obtained from animals if they are individually marked, and we recommend doing so.

In practice, the major problem we find with mark-recapture methods is defining the study area. This is not a problem if working on islands or estimating deer abundance within high fences. But this is a real problem when using live traps in a defined grid to mark-recapture groups such as small mammals. We recommend using the maximum daily movement (obtained from maximum distance between traps within the grid in which an individual was trapped on consecutive days) of the animals in question to define the limits outside the grid.

Jolly-Seber Method

The *Jolly-Seber method*, used for open populations, estimates population size, survival rates, and births based on a mark-recapture experiment. During each survey, all animals caught are marked, the identity of marked individuals is recorded, and all animals are then released. An estimate of population size is calculated from the simple relationship that population size is equal to the size of the marked population divided by the proportion of animals marked. The Jolly-Seber method is data intensive and has potentially difficult analyses, so practitioners are advised to use available computer software (e.g., program JOLLY, Pollock et al. 1990; program POPAN-5, Arnason and Schwarz 1999).

Schnabel Estimator

In situations where animals are continually being marked as surveys are conducted, there are several ways to analyze the data for a population estimate. A common way is to treat each recapture survey as a separate data set (using the total number marked at the time of the survey) to obtain multiple estimates and then calculate a mean estimate of population abundance using the Lincoln-Petersen estimator. Or, because the number of marked in the population affects the estimate, one could use only the data obtained from the final survey to obtain an estimate. If the former is used, equal weight is given to each survey when the last survey has more marked animals then the others; therefore, the last survey should provide a more accurate estimate. However, if only the last survey is used, then all the data available are not being used. To overcome this problem, Schnabel (1938) developed a weighted-average method to use all the available data without giving each survey an equal weight. The primary requirement for the *Schnabel estimator* is the same as for the Lincoln-Petersen estimator; the recapture sample has the same ratio of marked to non-marked as is found within the population.

Schumacher-Eschmeyer Estimator

Schumacher and Eschmeyer (1943) developed a variation of the Schnabel estimator, which is a variation of the Lincoln-Petersen estimator. Like the Schnabel estimator, the *Schumacher-Eschmeyer estimator* uses all the available data without giving each survey an equal weight. The assumption is the same as for the Lincoln-Petersen and Schnabel estimators in that the recapture sample has the same ratio of marked to non-marked as is found within the population (table 2.1).

Table 2.1. Hypothetical example of five days of trapping and marking mice

Day	Number trapped (n)	Number recaptured (m)	Number alive (A)	Total marked alive prior to date (M)	Mn	nM²	m
1	10	0	10	0	0	0	0
2	12	2	11	10	120	1,200	20
3	15	5	15	19	285	5,415	95
4	10	5	9	39	390	15,210	195
5	11	7	11	43	473	20,339	301
Totals		19			1,268	42,164	611

Note: Data are presented in a format suitable for estimation of population abundance using the Schnabel and Schumacher-Eschmeyer estimators. Note that only the first six columns are needed for the Schnabel estimator, whereas all eight columns are needed for the Schumacher-Eschmeyer estimator.

Quadrat Counts

Counts of plants or animals on areas of known size, such as quadrats, are among the oldest techniques in management to obtain population density data. These are routinely used by research organizations and natural resource agencies for habitat studies and management programs. We discuss two primary types of quadrat counts: (1) strip or cruise counts and (2) aerial surveys. These are widely implemented because of their simplicity and flexibility, but managers need to make important decisions during the design process. For instance, a manager who is going to sample a forest community to estimate the abundance of pine trees must first make two operational decisions: (1) What *size* quadrat should be used? (2) What *shape* quadrat is best? Long, thin quadrats are better than circular or square ones of the same area because of habitat heterogeneity. Long quadrats cross more patches since areas are never uniform.

Strip Counts

One of the most commonly used methods to measure density is *strip counts*. The counting unit is a strip or transect, which is merely a long, narrow rectangle of fixed area. Transects are randomly placed across the grain of the topography and landscape. Transect lines can be traversed on foot, horseback, truck, or boat or in a helicopter or airplane. The classic strip census uses a preset distance (e.g., 0.5-strip width) on each side of the transect line, and only those animals within this predefined distance are counted. It is assumed that all animals within the strip are counted with certainty. If these assumptions are valid, the population abundance can be estimated using any of the simple population estimators (Cochran 1977; Krebs 1998; Caughley and Sinclair 1994) for samples of equal area, samples of unequal area, or sampling proportional to size.

Aerial Surveys

Aerial surveys are a useful method for collecting data on a wide range of wildlife and wildlife habitats. These are particularly useful for larger wildlife species (deer) or wildlife sign (termite mounds, beaver dams) in areas visible from the air (less canopy cover to obscure observations), in the delineation of habitats (vegetation structure), and over large areas that make driving or walking impractical. There are three basic types of aerial surveys: (1) aerial transect sampling, (2) aerial quadrat sampling, and (3) aerial block sampling.

The most common type is *aerial transect sampling*. In this method, an aircraft flies in a straight line from one side of the census zone to the other at a fixed height above the ground. Streamers are attached to the wing struts of the plane so that the observer can project a strip on the ground. The observer records all relevant observations (animals, invasive tree groves) within that strip while flying along the transect.

In *aerial quadrat sampling*, the sample units are square or rectangular quadrats located at random within the census zone. The whole census zone can be set up as a checkerboard, and the quadrats to be searched can be determined by random numbers. Observers fly over the selected quadrat until they believe they have detected all individuals or collected all relevant information.

Aerial block sampling is similar to aerial quadrat sampling, except the sample units are blocks of land defined by physical features, such as rivers rather than a uniform overlaid grid (a checkerboard pattern). A sample of blocks to count can be chosen by locating random points in space and then counting the blocks in which a random point falls. Observers fly over the selected block until they believe they have detected all relevant data points.

Of these three methods, transect sampling is usually best when it can be applied. The sampling problems associated with aerial counts can be easily identified; for example, not all animals are seen in aerial counts, so a serious undercounting bias is almost always present. Additionally, it is difficult to individually identify animals, so double-counting is a concern when attempting to count all individuals in aerial quadrat or aerial block sampling.

Distance Methods

Distance sampling is a comprehensive approach encompassing study design, data collection, and statistical analyses (Buckland et al. 2001). When properly applied and when critical assumptions are met, distance data yield direct estimates of density (not simply abundance) that take probability of detection into account (Rosenstock et al. 2002). Density estimates can be expanded to yield estimates of total number of individuals by multiplying density by a known area of interest. Modern distance sampling is based on the observation that detection probabilities decrease with increasing distance from the observer (Burnham and Anderson 1984). In distance sampling, counts are assumed to be incomplete. Thus, the proportion of animals present that are actually seen must be estimated, and actual counts must be corrected by these detection probabilities. Perpendicular or radial distance data are used to estimate these detection probabilities. Data collection can be performed using either line transects or points. Distance sampling analyses can be quite sophisticated, but survey planning, including sampling design and estimates of sample size, can be performed within the Distance software program.

Line Transects

Similar to strip counts, *line transects* offer managers a simple method for measuring distance. Essentially, line transects are a boundless (plotless) version of the strip count method since the transect line replaces the long, thin quadrats. A line transect is searched, and each animal seen provides one measurement of the perpendicular distance to the transect line. Transect lines may be traversed on foot, on horseback, in a vehicle, or in a helicopter or airplane. Three measurements can be taken for each sighted individual: sighting distance, sighting angle, and perpendicular distance. Although this method is straightforward, several assumptions must be made to utilize the data: (1) animals directly on the transect line will never be missed (their detection probability = 1); (2) animals are fixed at the initial sighting position (they do not move before being detected, and none are counted twice); (3) distances and angles are measured exactly with no measurement error and no rounding errors; and (4) sightings of individual animals are independent events.

Point Counts

Point counts are typically used to estimate bird density. Using this method, an observer proceeds to a sample point. The observer then waits for a period of time prior to data collection to allow bird activity to return to normal after observer arrival (Reynolds, Scott, and Nussbaum 1980). The observer then detects birds (by both sight and sound) for a specified time period. Although the assumption is that all birds are detected at a point, typically point counts have been viewed as indices of abundance when standardized protocols are emphasized (Ralph, Sauer, and Droege 1995). Therefore, unless complete counts are certain, point counts should be considered as an index to relative density.

Time-Area Counts

Another point-based example of using distances to estimate the effective sample area of the counts is *time-area counts*. Plots are chosen at random in representative cover. Counters are stationed at each plot before sunrise. For 30 minutes the observer remains quiet and motionless while counting all species of interest, squirrels, for example, that come into view. After the count is completed, the average distance that a squirrel could be seen is determined by pacing. Using this determined distance as a radius, the observer computes the area of the circular plot that was watched. Since not more than three-fourths of the area can be observed without head movement, only three-fourths of the area can be considered, and only the squirrels seen in this area should be counted.

Point-Quarter Method

The *point-quarter method* is a classic method for sampling vegetation. Using this method, surveyors locate and describe the four trees nearest each corner of a section of land (e.g., 2.6 sq km, or 1 sq mi). Selected points from a sampling design are located in the field, and the area around the point is precisely divided into four (90-degree) quadrants (either perpendicular to the transect if point-transect sampling, or by compass bearing for random points). The distance from the point to the nearest target within each quadrant is measured, so four distances are obtained at each point. Population densities are then derived from these distances.

Sampling with Probability Proportional to Size

Large study areas are seldom homogeneous with respect to resources, species density, or detectability. Stratification is used to divide the large area into units of similar composition to reduce errors in data. As a result of this subdivision, the units to be sampled are often of unequal size. In these circumstances one can employ sampling with *probability proportional to size* (PPS), where the probability of a sample being selected is proportional to the size of the various units being sampled. The PPS method can be used with equal or unequal size sampling units. Sampling using the PPS method requires the density to be calculated for each sample, with the average density estimates to be calculated from all units sampled.

Exploited Population Techniques

These techniques were developed to estimate population size in exploited populations (e.g., hunted, fished). Three types of approaches to population estimation can be used with exploited populations: (1) change-in-ratio method, (2) Eberhardt's removal method, and (3) catch-effort method. The *change-in-ratio method* is useful in carefully studied hunted populations. Kelker (1940) showed that one could calculate a population estimate from the simple ratio when only one sex is hunted and the sex ratio before and after hunting, as well as the total kill, is known. One limitation of the method is that a substantial change in one of the organisms should occur (>20%) to obtain a reliable estimate.

Point-Quarter Method

Description: The point-quarter method is a tried-and-true way to sample vegetation. Fortunately, the method can also be used to determine wildlife density if the wildlife remains in stationary locations or can be measured prior to movement (e.g., nests, burrows). The broad applicability and relatively low cost and effort required to use this method have allowed it to remain a popular data-collection technique.

Scenario: Imagine you are tasked with determining the density of a particular tree species in a sample area. To do so, you will use the point-quarter method with sample points spread along a transect. Tree density is determined through the following calculation:

Density = $1 \text{ m}^2 \div$ (mean distance to tree)^2

Point-quarter method. Image by Joyce VanDeWater.

In the first sample point you determine the distances to the nearest study tree in each quadrant as follows:

Quadrant 1: 7 m	Quadrant 3: 3 m
Quadrant 2: 4 m	Quadrant 4: 6 m

Determine density of trees on this single sample point using the above distances. Use meters in all calculations. You can convert to trees/hectare using this simple conversion: 10,000 m²/ha. Keep in mind the following issues (for more detailed information, see Pierce, Lopez, and Silvy 2012):

1. This is the density of a single point. The study area would have multiple sample points.
2. This is a rough estimate without associated error estimates.

Eberhardt's removal method provides a simple population density estimation if an index of population size (e.g., roadside counts) can be made before and after the removal of a known number of individuals. This method does not require classification of individuals into *x*-types and *y*-types as does the change-in-ratio method, and there is no need to identify individuals.

The *catch-effort method* estimates population size by recording the decline in catch-per-unit effort (e.g., catch/day) with time. If catchability is constant and this decline is linear, regression methods can be used to estimate population size at the time when exploitation begins. Fisheries managers have used this method extensively, but the assumption of constant catchability is vital if the population estimates are to be accurate.

Emerging Technologies

Wildlife habitat management benefits from technological innovation, as do many other human endeavors. In fact, many of the emerging technologies in wildlife habitat management stem from advances in other areas, such as computing, spatial sciences, medicine, and aeronautics. The complex and varied requirements for effective management engender a willingness to embrace new technologies and the ability to alter those technologies to fit specific needs. These needs are as broad as the natural resources being managed and include areas such as advances in remote wildlife tracking, more complex and user-friendly analytical software, component miniaturization, emergence of drones in aerial reconnaissance and tracking, improving camera optics and battery life, declining costs for high-precision equipment such as cameras and global positioning units, new captive propagation techniques and technology, and emerging outreach technologies.

It is impossible to catalog all of them because they are simply too varied and are constantly evolving at the pace of technological change in our society. It is also difficult to predict which technologies will emerge; however, we can generally surmise that improved autonomous monitoring of wildlife and their associated habitats will continue to develop. The improvement in computing power and imaging technology is likely to positively impact remote camera, satellite, and wildlife tracking technology to significant degrees. We can also guess that future wildlife habitat managers will use emerging technologies in unexpected and innovative ways to confront new conservation issues. Here, we focus on a few of the emerging and improving technologies in wildlife habitat management and give some examples.

Drones

Drones have made their way into the public consciousness and lexicon relatively quickly. Various names are attached to these vehicles, but the differences are still under debate and unimportant for our purposes (Sandbrook 2015). Unmanned aerial vehicles are commonly mentioned in reference to military operations, commercial and private operations, and emerging airspace conflicts. These unmanned vehicles (generally aerial in this context) can be remotely controlled or preprogrammed to cover predetermined routes. Wildlife researchers have used small unmanned vehicles in niche situations in the past (e.g., to explore burrows), but increasingly drones are being adopted for widespread natural resources management for several reasons: improved battery life, improved command and control of the vehicles, decreased prices, improved cameras, and better analytical software. Drones also potentially offer alternatives to existing data-collection techniques when current methods are unavailable or impractical.

The potential exists for these vehicles to contribute to a number of different undertakings, from forest evaluations to vertebrate population monitoring. For example, researchers developed a small drone to monitor tropical forest biodiversity in parts of Indonesia where remote sensing capabilities (satellites) were limited (Koh and Wich 2012). Even this early prototype drone was able to fly autonomously for over 15 kilometers (9.3 mi). Lisein et al. (2013) focused on practical drone functionality when they analyzed several methods for estimating strip sample areas during wildlife surveys. The increasing practicality of drones to conduct these types of surveys can lower costs associated with hiring fixed-wing or rotary aircraft and allows more sampling. Challenges associated with range, battery life, airspace conflicts, and relevant laws remain important; but the usefulness of drones is only likely to increase into the future.

Remote Cameras

Remote cameras have become firmly entrenched in natural resources research and management. They provide data that researchers find impractical or impossible to collect in person. Fortunately, the technology for remote cameras has continued to improve, making cameras more reliable, portable, and accurate. Battery life has markedly improved, depending on demands and location, with units routinely left in the field for months at a time. High-capacity batteries (e.g., NiMH, lithium) are readily available, along with solar chargers and additional battery packs that can further increase deployment capabilities (Brown and Gehrt 2009). In addition to battery improvement, data storage capabilities are much better as the cameras have decreased in price and physical size.

Remote cameras are now able to accommodate large-capacity onboard and supplemental storage devices capable of holding thousands of pictures and videos (Newbery and Southwell 2009). Cameras have increased fields of view, shutter speeds, image resolution, and user operability even as sizes have decreased. For instance, Thompson et al. (2012) were able to place high-resolution cameras onto migrating caribou (*Rangifer tarandus*) to study behavior. This is the type of application that a researcher would be hard-pressed to accomplish in person. Researchers are also utilizing increased technology in remote uplinks and can often download data directly from the unit. This further minimizes the researcher's or manager's footprint in the study area.

One of the more visually stunning developments in remote camera technology has been thermal infrared cameras. Until recently, these units have been too large or far too expensive to use regularly. Thermal cameras have decreased in size and price, though they are by no means cheap. They have been used most successfully so far in detecting body heat of mammals and large birds. Thermal infrared cameras have been used to determine density of mammal species, detect diseased animals, and study thermoregulatory processes. As prices and size continue to decrease, these cameras will likely be used more extensively.

Wildlife Telemetry

The technology of tracking wildlife has benefited a great deal from advances in other realms of science, including reduction in component sizes, reduced costs, better sensors, improved satellite technology, and access to satellites (Millspaugh et al. 2012). Furthermore, the improved computing capabilities and software for analyzing wildlife and habitat use contribute a great deal to the ease of use of wildlife transmitters. Very high frequency (VHF) transmitters have been a staple of wildlife research and management for about half a century and remain critical for species located in high-cover areas that interfere with satellite-based systems (Gitzen et al. 2013). The first of these units were rather large and useful only to survey large animals like caribou. Even with relatively short battery life and transmission ranges, these early prototypes proved enormously useful for wildlife biology, including analyzing wildlife habitat use. Imagine the insights available to researchers once they were able to track migratory animals; cryptic, dangerous, or fragile species; and marine animals. Over the intervening decades wildlife transmitters have become common in wildlife biology, ranging from collar mounts to backpacks and implants.

In recent years the technology has expanded beyond the VHF technology used since the early days. It is not unusual for researchers and managers to use satellite-based systems such as global positioning system (GPS) tags or platform terminal transmitters (PTTs) to accurately track animals over large distances and independent of on-the-ground observers. These are sometimes integrated with communication networks (such as cellular networks) to improve data collection (Millspaugh et al. 2012). Even light levels are used to track animals, because a global location sensing (GLS) device allows researchers to retrieve tags that record light levels experienced by the animals. This information is then used to determine location. Technology has improved to the point that GPS, PTT, and GLS systems can be used on small migratory birds (Bridge et al. 2011). These technologies have also been used in concert with biological sensors to monitor both animal conditions (e.g., heart rate, body temperature) and environmental data (e.g., ambient temperature, salinity; Wilson et al. 2015). As all of these technologies continue to diversify, shrink in size, improve in battery life or energy use, and decline in costs, they promise to become even more important in wildlife biology.

Geospatial Technologies

Geospatial technologies are relatively common in wildlife habitat management with the entire field of spatial ecology dedicated to its use. Skidmore et al. (2011) described spatial ecology as focusing on the role of space and time in ecological processes at multiple scales. The emerging field is critical to analyzing current wildlife habitat realities and determining future management strategies. Fortunately, spatial data and techniques have improved markedly. Satellite and aerial imagery is continuously improving in availability and resolution. Researchers and managers in many urban global north countries can expect high-quality imagery at multiple scales and of multiple data sets to be readily available. This availability declines significantly in rural or global south regions of world, however.

Spatial ecology is a broad field with many resources at its disposal, including satellite remote sensing, GPS, and geographic information systems. These have all been used for a variety of purposes, such as assessing biodiversity and habitat fragmentation, monitoring landscape dynamics, creating habitat models, and planning conservation priorities (Yadav, Sarma, and Dookia 2013). Techniques are emerging that take advantage of powerful technologies. For instance, LIDAR is a laser-based remote sensing technology that can provide highly detailed, three-dimensional landscape data that managers can use for a variety of habitat management purposes (Merrick, Koprowski, and Wilcox 2013). More recently, LIDAR has been used to sample vegetation biomass as an alternative to conventional field sampling. The improvement in availability of these resources and their relatively reasonable costs make them indispensable components of current and future wildlife habitat management.

Summary

I. Why Sample?

   A. Vegetation, wildlife populations, and land are the building blocks of wildlife management.

   B. Understanding these three components helps wildlife managers assess wildlife and associated habitats.

II. Equipment and Techniques

   A. Managers and research biologists use a variety of equipment and techniques to sample and measure vegetation in a multitude of different aquatic and terrestrial plant community and habitat types.

   B. Terrestrial vegetation assessment is fairly straightforward, but sampling and measurement techniques vary considerably among grassland, shrubland, and woodland habitat types.

   C. Effective managers understand how and why to apply different equipment and techniques.

III. Vegetation Sampling

   A. Vegetation sampling and measurement are generally conducted within vegetation patches or field-size units. This concept is critical to habitat management because wildlife reacts to disturbance in a variety of ways.

   B. Landscale-level assessments of vegetation usually couple geographic information system techniques with three data sources:

      1. Satellite imagery
      2. Aerial imagery
      3. Video photography

   C. Recent advances in computer capabilities have enabled managers and researchers to work with larger and more complex data sets to assess vegetation characteristics. These capabilities also enable the integration of vegetation data with other data sets, such as the following:

   1. Animal population demographics
   2. Weather
   3. Topography
   4. Soils

IV. Animal Sampling

A. General animal abundance estimation utilizes a variety of methods depending on data requirements and circumstances but falls into three areas:

   1. Total counts
   2. Indices
   3. Population estimates

B. Resource managers can choose from these well-studied tools as situations require (e.g., point counts for bird abundance estimation).

C. Development of computational capabilities and analytical techniques has allowed even more avenues of discovery.

D. These new computational methods require meeting unique sets of assumptions (e.g., closed vs. open populations), thus placing ever more emphasis on the experience and knowledge of the investigators.

V. Emerging Technologies

A. Wildlife habitat management and related research benefit from technological innovation occurring in society.

B. These technologies are varied, but we focus on just a few important examples:

   1. Drones
   2. Remote cameras
   3. Wildlife telemetry
   4. Geospatial technologies

C. The improvement in these and other technologies will provide wildlife habitat managers with new and powerful tools.

Literature Cited

Anderson, D. R. 2001. The need to get the basics right in wildlife field studies. *Wildlife Society Bulletin* 29:1294–97.

Arnason, A. N., and C. J. Schwarz. 1999. Using POPAN-5 to analyze banding data. *Bird Study* 46:S157–S68.

Bajzak, D., and J. F. Piatt. 1990. Computer-aided procedure for counting water-fowl on aerial photographs. *Wildlife Society Bulletin* 18:125–29.

Bitterlich, W. 1948. Die winkelzahlprobe. *Allgemeine Forst-und Holzwirtschaft-lich Zeitung* 59:4–5.

Bonham, C. D. 1989. *Measurements for terrestrial vegetation*. New York: John Wiley and Sons.

Bontaites, K. M., K. A. Gustafson, and R. Makin. 2000. A Gasaway-type moose survey in New Hampshire using infrared thermal imagery: Preliminary results. *Alces* 36:69–75.

Bridge, E. S., K. Thorup, M. S. Bowlin, P. B. Chilson, R. H. Diehl, R. W. Fléron, P. Hartl, R. Kays, J. F. Kelly, W. D. Robinson, and M. Wikelski. 2011. Technology on the move: Recent and forthcoming innovations for tracking migratory birds. *BioScience* 61:689–98.

Brown, J., and S. Gehrt. 2009. *The basics of using remote cameras to monitor wildlife*. Ohio State University Extension Fact Sheet W-21-09. Columbus: Ohio State University.

Buckland, S. T., D. R. Anderson, K. P. Burnham, J. L. Laake, D. L. Borchers, and L. Thomas. 2001. *Introduction to distance sampling*: *Estimating abundance of biological populations*. Oxford: Oxford University Press.

Burnham, K. P., and D. R. Anderson. 1984. The need for distance data in transect counts. *Journal of Wildlife Management* 48:1248–54.

Caughley, G. 1977. *Analysis of vertebrate populations*. New York: John Wiley and Sons.

Caughley, G., and A. R. E. Sinclair. 1994. *Wildlife ecology and management*. Boston: Blackwell Scientific Publications.

Cochran, W. G. 1977. *Sampling techniques*. 3rd ed. New York: John Wiley and Sons.

Daubenmire, R. F. 1968. *Plant communities*: *A textbook of plant synecology*. New York: Harper and Row.

Gill, R. M. A., M. L. Thomas, and D. Stocker. 1997. The use of portable thermal imaging for estimating deer population density in forest habitats. *Journal of Applied Ecology* 34:1273–86.

Gitzen, R. A., J. L. Belant, J. J. Millspaugh, S. T. Wong, A. J. Hearn, and J. Ross. 2013. Effective use of radiotelemetry for studying tropical carnivores. *Raffles Bulletin of Zoology* 28:67–83.

Goodrum, P. D., V. H. Reid, and C. E. Boyd. 1971. Acorn yields, characteristics, and management criteria of oaks for wildlife. *Journal of Wildlife Management* 35:520–32.

Greig-Smith, P. 1964. *Quantitative plant ecology*. New York: Plenum Press.

Gysel, L., and L. J. Lyon. 1980. Habitat analysis and evaluation. In *Wildlife management techniques manual*, 4th ed., edited by S. D. Schemnitz, 305–27. Bethesda, MD: Wildlife Society.

Haramis, G. M., J. R. Goldsberry, D. G. McAuley, and E. L. Derleth. 1985. An aerial photographic census of Chesapeake Bay and North Carolina canvasbacks. *Journal of Wildlife Management* 49:449–54.

Higgins, K. F., and W. T. Barker. 1982. *Changes in vegetation structure in seeded nesting cover in the prairie pothole region.* US Department of the Interior, Fish and Wildlife Service Special Scientific Report 242. Washington, DC: US Fish and Wildlife Service.

Kelker, G. H. 1940. Estimating deer population by a differential hunting loss in the sexes. *Proceedings of the Utah Academy of Sciences, Arts and Letters* 17:65–69.

Koh, L. P., and S. A. Wich. 2012. Dawn of drone ecology: Low-cost autonomous aerial vehicles for conservation. *Tropical Conservation Science* 5:121–32.

Krebs, C. J. 1998. *Ecological methodology.* New York: Harper and Row.

———. 1999. *Ecological methodology.* 2nd ed. New York: Harper and Row.

Lisein, J., J. Linchant, P. Lejeune, P. Bouché, and C. Vermeulen. 2013. Aerial surveys using unmanned aerial system (UAS): Comparison of different methods for estimating the surface area of sampling strips. *Tropical Conservation Science* 6:506–20.

McCullough, D. R. 1979. *The George Reserve deer herd: Population ecology of a K-selected species.* Ann Arbor: University of Michigan Press.

Merrick, M. J., J. L. Koprowski, and C. Wilcox. 2013. *Into the third dimension: Benefits of incorporating LiDAR data in wildlife habitat models.* USDA Forest Service Proceedings RMRS-P-67. Fort Collins, CO: US Department of Agriculture, Forest Service, Rocky Mountain Research Station.

Millspaugh, J. J., D. C. Kesler, R. W. Kays, R. A. Gitzen, J. H. Shultz, C. T. Rota, C. M. Bodinof, J. L. Belant, and B. J. Keller. 2012. Wildlife radiotelemetry and remote monitoring. In *The wildlife techniques manual: Management,* 7th ed., edited by N. J. Silvy, 258–83. Bethesda, MD: Wildlife Society.

Mueller-Dombois, D., and H. Ellenberg. 1974. *Aims and methods of vegetation ecology.* New York: John Wiley and Sons.

Newbery, K. B., and C. Southwell. 2009. An automated camera system for remote monitoring in polar environments. *Cold Regions Science and Technology* 55:47–51.

Nudds, T. D. 1977. Quantifying the vegetative structure of wildlife cover. *Wildlife Society Bulletin* 5:113–17.

Oldemeyer, J. L., and W. L. Regelin. 1980. Comparison of 9 methods for estimating density of shrubs and saplings in Alaska. *Journal of Wildlife Management* 44:662–66.

Olenicki, T. 2001. Ground-based radiometers, real-time GPS receivers, and laser rangefinders—new techniques for estimating vegetation parameters and animal use sites. *Intermountain Journal of Sciences* 6:384–85.

Owensby, C. E. 1973. Modified step-point system for botanical composition and basal cover estimates. *Journal of Range Management* 26:302–3.

Parker, H. D., Jr., and R. S. Driscoll. 1972. An experiment in deer detection by thermal scanning. *Journal of Range Management* 25:480–81.

Pierce, B. L., R. R. Lopez, and N. J. Silvy. 2012. Estimating animal abundance. In *The wildlife techniques manual: Management,* 7th ed., edited by N. J. Silvy, 284–310. Bethesda, MD: Wildlife Society.

Pollock, K. H., J. D. Nichols, C. Brownie, and J. E. Hines. 1990. Statistical inference for capture-recapture experiments. *Wildlife Monographs* 107:3–97.

Ralph, C. J., J. R. Sauer, and S. Droege. 1995. *Monitoring bird populations by point counts.* US Department of Agriculture, Forest Service, General Technical

Report PSW-GTR-149. Albany, CA: Pacific Southwest Research Station, Forest Service, US Department of Agriculture.

Raven, P. H., R. F. Evert, and S. E. Eichhorn. 1986. *Biology of plants*. 4th ed. New York: Worth Publishers.

Reynolds, R. T., J. M. Scott, and R. A. Nussbaum. 1980. A variable circular-plot method for estimating bird numbers. *Condor* 82:309–13.

Robel, R. J., J. N. Briggs, A. D. Dayton, and L. C. Hulbert. 1970. Relationships between visual obstruction measurements and weight of grassland vegetation. *Journal of Range Management* 23:295–97.

Rosenstock, S. S., D. R. Anderson, K. M. Giesen, T. Leukering, and M. F. Carter. 2002. Landbird counting techniques: Current practices and an alternative. *Auk* 119:46–53.

Sandbrook, C. 2015. The social implications of using drones for biodiversity conservation. *Ambio* 4:636–47.

Schnabel, Z. E. 1938. The estimation of the total fish population of a lake. *American Mathematical Monthly* 45:348–52.

Schumacher, F. X., and R. W. Eschmeyer. 1943. The estimate of fish population in lakes or ponds. *Journal of the Tennessee Academy of Sciences* 18:228–49.

Schupp, E. W. 1990. Annual variation in seedfall, postdispersal predation, and recruitment of a Neotropical tree. *Ecology* 71:504–15.

Shafer, E. L., Jr. 1963. The twig-count method for measuring hardwood deer browse. *Journal of Wildlife Management* 27:428–37.

Skidmore, A. K., J. Franklin, T. P. Dawson, and Petter Pilesjö. 2011. Geospatial tools address emerging issues in spatial ecology: A review and commentary on the special issue. *International Journal of Geographical Information Science* 25:337–65.

Thompson, I. D., M. Bakhtiari, A. R. Rodgers, J. A. Baker, J. M. Fryxell, and E. Iwachewski. 2012. Application of a high-resolution animal-borne remote video camera with global positioning for wildlife study: Observations on the secret lives of woodland caribou. *Wildlife Society Bulletin* 36:365–70.

Thompson, R. L., and B. S. McGinnes. 1963. A comparison of eight types of mast traps. *Journal of Forestry* 61:679–80.

Undersander, D. J., B. Albert, P. Porter, and A. Crossley. 1992. *Pastures for profit: A hands on guide to rotational grazing*. University of Wisconsin Extension Service No. A3529. Madison: University of Wisconsin Extension Service.

Whittaker, R. H. 1965. Branch dimensions and estimation of branch production. *Ecology* 46:365–70.

Wiggers, E. P., and S. F. Beckerman. 1993. Use of thermal infrared sensing to survey white-tailed deer populations. *Wildlife Society Bulletin* 21:263–68.

Wight, H. M. 1939. *Field and laboratory technic in wildlife management*. Ann Arbor: University of Michigan Press.

Wilson, A. D. M., M. Wikelski, R. P. Wilson, and S. J. Cooke. 2015. Utility of biological sensor tags in animal conservation. *Conservation Biology* 29:1065–75.

Yadav, P. K., K. Sarma, and S. Dookia. 2013. The review of biodiversity and conservation study in India using geospatial technology. *International Journal of Remote Sensing and GIS* 2:1–10.

3   Analysis of Wildlife Habitat

We lay the foundation for development of research and wildlife management plans by discussing what underlies scientific knowledge: methods of pursuing science. Clearly the research conducted, and the management that is built on that research, will be less than optimal if based on nonrigorous study. For example, research papers and agency reports often contain a summary and review of previous research on the topic of interest, in which the author(s) simply repeats the conclusions of past studies without also providing any information on associated sample sizes, number of study areas, and so forth. What does this mean to managers? It means that managers need to clearly understand what constitutes rigorous science, including how to design studies, collect and analyze data, present and interpret results, and turn their findings into practical management applications. Managers must have a fundamental understanding of scientific process to judge the quality of research and to implement recommended management actions.

Science and Reliable Knowledge

Although solid science forms the basis of solid management, the jump from science to management applications is as difficult as the design and analysis of scientific study. The goals of the research must be developed such that the results can translate into meaningful management. For example, increasing songbird abundance is often a management goal for private landowners attempting to qualify for wildlife habitat state-tax valuations in Texas. Although it is interesting to determine the rate of survivorship in adult songbirds, there is usually little we can do to change it. There are more readily available management options, however, to influence the survivorship of eggs and nestlings. Thus, while we would never advise against studying adult survivorship, managers would not want to ignore breeding success if the goal was to increase the population size of a species (especially those that are rare).

The key to transferring scientific knowledge to management application begins with development of the management plan. In this chapter, we discuss developing and conducting scientific research. We discuss the development of management plans in later chapters. For the management plans, we draw heavily from our direct experience with student-developed plans. We also discuss the related topics of professional writing and ethics in writing and provide guidance on how to avoid many of the most common writing errors.

A central goal of scientific methods is to provide knowledge that is repeatable. Gaining confidence in the knowledge we acquire allows us to advance our understanding of wildlife and its habitats and moves us toward an ability to predict the future consequences of natural and human-induced changes in the environment. We must avoid making management decisions based on our biases that derive from poor information. The discipline of ecology in general has failed to provide reliable knowledge not because it is inherently flawed but because its practitioners have failed to treat it as rigorous science (Romesburg 1981; Peters 1991).

Study Design

There exist basic sets of principles for how to implement a scientific study (study design) that are common to all disciplines. Even those who never conduct a study will continually be required to evaluate the results of, and interpret the quality of, work conducted by other people. This is a critical skill for natural resource managers and landowners. As implementers of management recommendations they must be able tell good from bad. Researchers and managers must understand how to collect and analyze data even if they never do the research themselves.

Scientific Methods

Development of the study goals follows from the statement of the problem. For example, what is causing most of the young produced by late-breeding females to die before winter? A review of the scientific literature, consultations with experts, and any existing field observations will lead to various hypotheses concerning the underlying reasons for excessive mortality of young (developing null hypotheses that generally state sample observations are due to chance). When stating study goals, the researcher must also establish the spatial and temporal applicability of the results. In our current example of mortality of young, if the work is restricted to a single study area and time frame (number of years), then the results will be unlikely to have broad applicability.

When reading and reviewing the scientific literature, the researcher will be confronted with an often bewildering array of manipulations of data and associated statistical analyses. Statisticians have established numerous requirements that must be met for the proper application of specific statistical analyses.

Much of the fascination with forcing biological data into a specific statistical box comes from a misunderstanding of the term "significant." We use "significant" in a host of biological, statistical, and social applications. When ecologists refer to *biologically significant*, they mean that there is enough difference or a strong enough relationship to cause us to think that particular data have biological meaning. Because we would never expect that exactly no difference exists between two biological entities, we need to specify how much difference we think would matter to the entities of interest (Cherry 1998; Johnson 1999). For example, given a large enough sample size, a difference in body mass of 10% between adult and juvenile might be statistically significant (e.g., $P < 0.05$); but does this difference result in differential overwinter survival?

Types of Studies

Green (1979) established a framework for categorizing studies as either "optimal" or "suboptimal" (fig. 3.1). Although he developed his framework in the context of environmental impact assessment, Green's categorization of studies applies to all research and is a convenient way to begin assessing the strength of a study a manager might be reviewing. The prerequisites for an *optimal study design* were that the treatment (or environmental impact, where *impact* applies to any natural or human-induced action) must not have occurred so that before-impact baseline data can provide a temporal control for comparing with post-impact data. Thus, nonimpact controls must be available (fig. 3.1, sequence 1). The classic manipulative experiment is potentially optimal because it alone offers replication and randomization. If these criteria for an optimal design are met, a manager should be able to test the null hypothesis that any change in the impacted area does not differ from the control and relate to the impact any demonstrated change unique to the impacted area (which allows the manager to separate nonimpact effects caused by naturally occurring variation).

In reality, however, managers and ecologists often (or perhaps most often) cannot meet the criteria for use of an optimal design, which are most frequently applicable in controlled settings, often in the laboratory or a very controlled (e.g., plots, pens, small study areas) field situation. Optimal designs are difficult to apply because we are usually interested in exploring naturally occurring ecological phenomena. Furthermore, environmental change or impacts often occur unexpectedly, or control

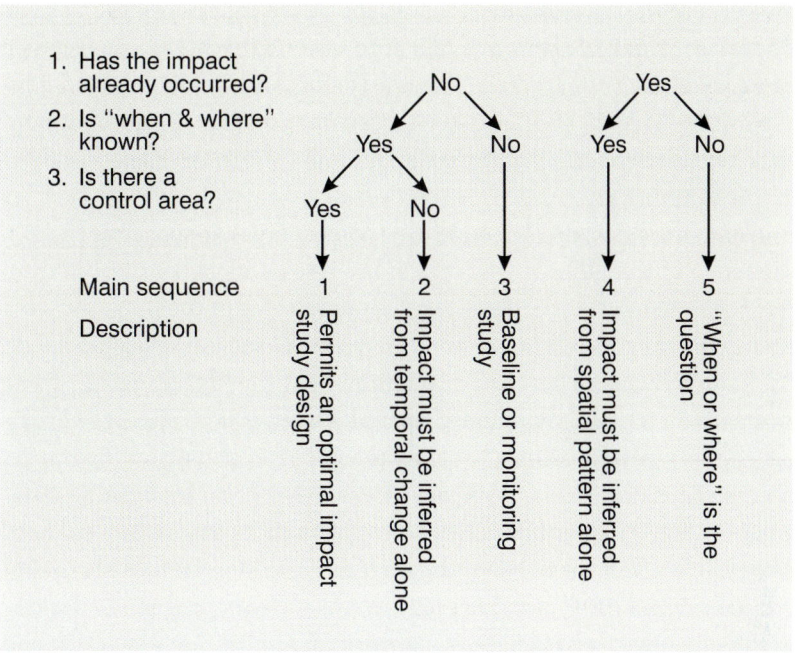

Figure 3.1. Framework for categorizing environmental studies as "optimal" or "suboptimal." From Green (1979, fig. 3.4).

Population Problems

Description: Many of the wildlife population issues we face are wildlife habitat issues. We often manage these wildlife populations by modifying existing habitat or creating new habitat. The strategies for confronting these issues are complex and rarely have silver-bullet solutions.

Scenario: Locate an article on a "population problem," such as disruption of migration or dispersal, small population size, isolated populations, source-sink or ecological trap, and briefly review the issue (about one-half page) and then propose a solution that might overcome (restore) the situation (about one and one-half pages). This is not a critique of the paper but a suggestion about how to fix the issue. The article itself might have such a solution; if so, briefly outline the suggestion and offer agreement, suggest changes, or propose a different approach. There is unlikely to be a "correct" answer, although some answers might be more realistic than others. Make sure to provide the citation for the paper you read, using this standard format:

Morrison, M. L., K. S. Smallwood, and L. S. Hall. 2003. Creating habitat through plant relocation: Lessons from valley elderberry longhorn beetle mitigation. *Ecological Restoration* 21:95–100.

areas are unfeasible, limited in availability, or logistically impossible. Therefore, much of our work falls into Green's *suboptimal study design* category because (1) treated (impacted) sites are usually selected non-randomly, (2) replication of sites is often not possible, (3) pretreatment data are usually scanty or nonexistent, and (4) control areas are difficult to establish.

Under the suboptimal design, if no control areas are possible (fig. 3.1, sequence 2), then the significance of the impact must be inferred from temporal changes alone. If the location and timing of the impact are not known (it is expected but cannot be planned, such as a fire or flood), the study becomes a baseline or monitoring study (fig. 3.1, sequence 3). If the study is properly planned spatially, then it is likely that nonimpact areas will be available to serve as controls if and when the impact occurs. Thus, the field of "impact assessment" provides specific guidance on how to quantify treatment effects in as rigorous a manner as possible given these suboptimal situations; Morrison et al. (2008) and Morrison (2009) provide more detail on developing study designs within an impact assessment framework for application to wildlife and restoration ecology.

Impacts often occur, however, when no preplanning by a scientist or land manager is reasonably possible. This common situation (fig. 3.1, sequence 4) means that impact effects must be inferred from among areas differing in the degree of impact. Finally, situations do occur (fig. 3.1, sequence 5) where an impact is known to have occurred, but the time and location are uncertain (e.g., the discovery of a toxin in a plant or soil).

Graphics and Descriptive Statistics Figures

The first step all managers and researchers should take after completing data collection is to visually examine the data. Unfortunately, we have found that visual representation of data (graphical displays) has become a lost practice. There are several reasons why the graphical step of data analysis is skipped, including a rush to see if statistical differences between study groups were evident and the related unhealthy reliance on a statistical test to determine biological validity. There are, however, several key reasons why data should be plotted or otherwise visualized prior to conducting any statistical tests: (1) Plotting provides a clear understanding of the distribution of the data, including identification of nonnormal (a most likely outcome) distributions of data. (2) Plotting identifies what appear to be outliers, data points that are well separated in space or time from the majority of the data. (3) Plotting gives an initial understanding of apparent relationships between variables, such as between animal numbers and canopy cover. The graphics in fig. 3.2 illustrate each of these major points.

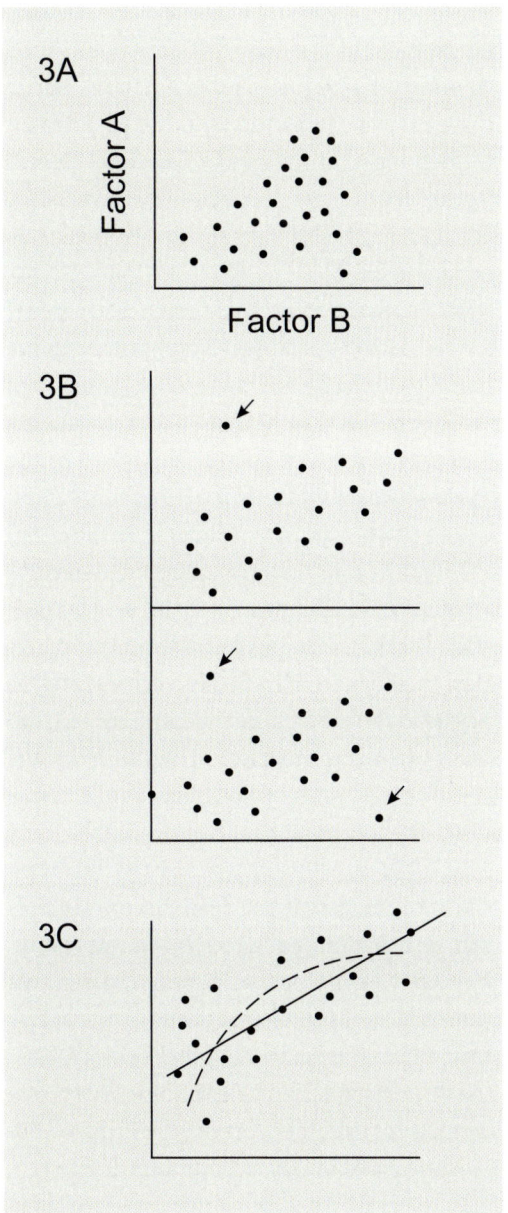

Figure 3.2. Hypothetical relationships between environmental factors. (A) Potential nonnormal distribution of data with a skewed distribution. (B) Potential linear relationships with one (top panel) and two (bottom panel) outliers. (C) Potential nonlinear relationship fitted with a curvilinear (dashed) and linear (solid) line. Image by Joyce VanDeWater.

In figure 3.2A there is a clear skewing of the data to the right of the x-axis, indicating that the data do not follow a normal distribution, which is a biological phenomenon worth preserving if the sample size is adequate. We say "worth preserving" because many researchers will first attempt to "normalize" the data by applying a mathematical transformation to the data. Although beyond the scope of this book, such transformations are done to meet the requirements for parametric tests and include converting the data to a log scale. Making such transformations usually obscures the biological pattern of the data. Nevertheless, viewing the data in graphical form is an essential part of the biological interpretation of the data regardless of subsequent decisions on transformations and statistical tests.

In figure 3.2B, top panel, there appears to be a positive relationship, perhaps linear, between the two plotted variables. Note, however, the single data point (arrow) that lies outside the swarm of other data points. Data points that are well separated from most other points are usually termed *outliers* because they are found outside the range of the other data points. In many cases there could be multiple outliers, such as depicted in figure 3.2B, bottom panel (arrows). Multiple outliers can often influence the outcome of a statistical test, for example, making a statistical test significant or not significant (e.g., $P < 0.05$ or $P > 0.05$, respectively) depending on their inclusion or exclusion from the analysis. Thus, how single or multiple outliers are handled is a critical step in data manipulation. In general, there should be a clear reason for excluding an outlier, such as a known recording mistake or malfunctioning equipment. A biological reason for excluding an outlier could include data collected under very unusual conditions that are not likely to be regularly repeated, such as extremely unusual weather or the low overflight of an aircraft. We are not implying that the response of animals to unusual weather, an aircraft, or other disturbances is not important. However, if a study is not specifically sampling the response of animals to such unusual conditions, then all the researcher is doing by including the data collected under those conditions is obscuring the results for the main study goal. In all cases the researcher should explain in reports and publications why any data points were excluded, especially if they were not the result of a known recording error or equipment malfunction.

Figure 3.2C depicts a clearly nonlinear relationship; we can generally describe the scatter of data points as curvilinear. The biological interpretation is one of a rapidly increasing change in the two variables up to a certain point, followed by a gradual decrease of the slope of the curve; that is, it flattens out. There are many situations in animal ecology that result in a curvilinear relationship; for example, the foraging rate of an animal is likely to increase as prey becomes more abundant. But because of physical limitations, an animal can locate, capture, and consume prey only at some maximum rate; thus, its foraging rate must asymptote (curve approaches zero). Likewise, the abundance of a local population of a species can increase rapidly until the available space (e.g., isolated patch of habitat) starts to become saturated; the rate of increase will likely slow as saturation nears because of intraspecific interactions.

As we see in figure 3.2C, the researcher can fit a line to the curved relationship (the actual statistical means of doing curve fitting is beyond the scope of this book). Clearly, forcing a linear relationship to the scatter of data points in figure 3.2C would only obscure the curved relationship and, when analyzed statistically, result in a false biological model. Here again we witness the clear necessity of viewing the data rather than just importing the data points into a database and applying a popular statistical analysis. We are in no way implying that statistical analyses are inappropriate. To the contrary, we are recommending that analyses be applied that are appropriate to the phenomenon at hand.

Tables

Presentation of data and results is often an underappreciated but critical component for research and management. Well-structured tables simplify comprehension and organize often complex information for readers. Tables are designed to make information simpler to understand, but not all situations demand tabular structure. Small sets of information are often best left in sentence structure with parenthetically organized data as necessary. A general rule of thumb is to avoid tables with fewer than three columns or rows of data/results. If the information is simple enough to populate such a small table, then it should be either incorporated into the text or its importance reevaluated.

When complex results or information proves too confusing or unwieldy for placement in the text, tables can provide an organized structure for presentation. A basic table is constructed of clearly and concisely labeled columns and rows of individual cells containing data. Columns are often used to separate data into groups or classes with rows used as summary areas (Collier and Schwertner 2012). This structure generally includes at a minimum a table legend or caption that must include organisms or systems described in the table, location and dates of the data collection, and brief description of table data. Tables must be referenced with an in-text citation and clearly support an idea or statement. Tables should not serve as a dumping ground for mass data or results or only vaguely support ideas in the text (e.g., avoid statements such as "see table 1 for relevant data"). Tables should be understandable without reference to additional text. In other words, the table must make sense if viewed as a stand-alone document. Additionally, a table should not simply repeat information already stated in the text. In this vein, authors should try to keep the structure as simple as possible to improve clarity and focus.

Table 3.1 demonstrates the importance of these concepts (note that we slightly changed the format of the caption and table from that in the original publication). Despite the complex nature of the analyses, LaFever et al. created a simple table with enough information in the caption to enable understanding without the supporting article. The caption includes geographic area, research focus, species of interest, and scenarios. This is supplemented by a brief footnote that clarifies some of the data formatting. The headings are clear without multiple footnotes, acronyms, or abbreviations. Thus, tables are incredibly useful for distilling complex data/results into a more understandable format but must follow a set of general guidelines to maximize their benefits.

Descriptive Statistics

Statistics are a useful tool to discern what is happening in a system, but they can be overwhelming in complexity. Fortunately, we often need only simple statistics to get a general understanding of what is happening and guide future questions about the system. Thus, we limit this discussion to statistics that any managers can easily employ in their own system of interest. As the name implies, *descriptive statistics* are used to describe the primary characteristics of a set of data in quantitative terms. Descriptive statistics quantitatively summarize a data set rather than support infer-

Table 3.1. Lower Keys marsh rabbit habitat as related to scenarios of future sea-level rise

Sea-level rise scenario	Migration		No migration	
	Abandonment	Protection	Abandonment	Protection
Current conditions	1,172	1,172	1,172	1,172
Low	729 (1.5)	496 (2.5)	871 (1.5)	594 (2)
Medium	307 (2.5)	123 (4)	210 (4)	119 (5)
High	94 (3.5)	40 (3)	70 (3)	29 (4)

Source: LaFever (2006).
Note: Total area (hectares) of potential Lower Keys marsh rabbit habitat on Big Pine Key, Boca Chica Key, and the Saddlebunch/Sugarloaf Keys under scenarios of future sea-level rise, migration or no migration of vegetation upslope, and protection or abandonment of developed areas. Numbers in parentheses indicate the relative (proportional) decrease in habitat between sea-level-rise scenarios.

ences about the population that the data are thought to represent. It is critical to understand that these statistical methods are starting points for more detailed analyses and are susceptible to biases (e.g., outliers can greatly impact means). For instance, *mean* (average of observed values) and *median* (midpoint of observed values) are two common ways of describing the midpoint of continuous data. The *variance* of the data describes its spread around the mean. The manager can also establish confidence intervals (upper and lower bounds for the data range), which are data-derived intervals that likely would contain the unknown population parameter (Collier and Schwertner 2012).

When would a manager need such measurements? Suppose that we wanted to determine the basic current weight for white-tailed deer in a watershed. We would first report and describe the mean and standard deviation (measure of variability) of body weight of each sex and age class (e.g., adult males and females) for the sample of animals we gathered to provide a descriptive sense of the usual weights and their variation. We could then gather these data for multiple years to determine trends or conduct more complex analyses to better understand relationships between weight and other variables (e.g., precipitation, forest structure). If we look at a basic management example, we can see how some of these ideas might work in the real world. Suppose that a manager is interested in general health of a white-tailed deer herd on a wildlife management area and begins examining weight (table 3.2, an exception to the three-column/row rule mentioned earlier because clarity is increased by putting the example in table form). The small data set shown in table 3.2 shows the weights of both adult male and female white-tailed deer during a single season of captures.

Table 3.2. Average weights of adult male and female white-tailed deer

Year	Average weight (lb)
2005	120
2006	122
2007	125
2008	124
2009	129
2010	132
2011	135
2012	137

Summary

I. Science and Reliable Knowledge

A. Gaining knowledge and developing effective management require gathering rigorous and reliable information.

B. Effective management plans include these key points:

1. Provide repeatable data that builds confidence in knowledge gained.
2. Avoid biases in the collection and interpretation of data.
3. Proper study design so that an adequate amount and type of data are available for analysis.
4. Appropriate presentation and interpretation of data.
5. Translation and implementation of results into practical applications.

C. A central goal of scientific methods is to provide knowledge that is repeatable.

D. Gaining confidence in the knowledge we gain allows us to advance our understanding of wildlife and its habitats.

E. Reliable knowledge helps minimize making management decisions based on our biases.

F. The discipline of ecology in general has failed to provide reliable knowledge because its practitioners have failed to treat it as rigorous science.

II. Study Design

A. Scientific Methods

1. Implementing a scientific study requires a clear statement of goals and objectives.
2. A review of the scientific literature, consultations with experts, and any existing field observations will lead to the statement of potential hypotheses.
3. Central issues of spatial and temporal replication of the study largely determine the application of the study.
4. Understanding the difference between biological and statistical significance is a core issue in study design and interpretation.

B. Types of Studies

1. Differences between optimal and suboptimal study designs must be clearly understood prior to study initiation.
2. Suboptimal study designs, including various categories of impact assessment, have wide applicability in wildlife science but are difficult to properly implement.

III. Graphics and Descriptive Statistics Figures

    A. The critical first step in data interpretation is visual graphing and understanding the information gathered rather than rushing into statistical tests.

1. Plotting provides a clear understanding of the distribution of the data.
2. Plotting helps identify nonnormal (a most likely outcome) distributions of data.
3. Plotting identifies what appear to be outliers.
4. Plotting gives an initial understanding of apparent relationships between variables.

    B. Well-structured tables simplify comprehension and organize often complex information.

1. Tables are designed to make information simpler to understand, but not all situations demand tabular structure.
2. Tables should stand alone and be easy to understand.

    C. Descriptive statistics are used to describe the primary characteristics of a set of data in quantitative terms.

1. These general statistical methods are starting points for more detailed analyses as indicated by the initial results.
2. Simple statistics, such as confidence intervals, assist managers in the interpretation of population statistics.

Literature Cited

Chan-McLeod, A. C. A. 2003. Factors affecting the permeability of clearcuts to red-legged frogs. *Journal of Wildlife Management* 67 (4): 663–71.

Chen, J., J. F. Franklin, and T. A. Spies. 1992. Vegetation responses to edge environments in old-growth Douglas-fir forests. *Ecological Applications* 2 (4): 387–96.

———. 1993. Contrasting microclimates among clearcut, edge, and interior of old-growth Douglas-fir forest. *Agricultural and Forest Meteorology* 63:219–37.

———. 1995. Growing-season microclimatic gradients from clearcut edges into old-growth Douglas-fir forests. *Ecological Applications* 5 (1): 74–86.

Cherry, S. 1998. Statistical tests in publications of the Wildlife Society. *Wildlife Society Bulletin* 26:947–53.

Collier, B. A., and T. W. Schwertner. 2012. Management and analysis of wildlife biology data. In *The wildlife techniques manual*, vol. 1, 7th ed., edited by N. J. Silvy, 41–63. Baltimore: Johns Hopkins University Press.

Green, R. H. 1979. *Sampling design and statistical methods for environmental biologists*. New York: John Wiley and Sons.

Haddad, N. M., D. R. Bowne, A. Cunningham, B. J. Danielson, D. J. Levey, S. Sargent, and T. Spira. 2003. Corridor use by diverse taxa. *Ecology* 84 (3): 609–15.

Johnson, D. H. 1999. The insignificance of statistical significance testing. *Journal of Wildlife Management* 63:763–72.

LaFever, D. H. 2006. Population modeling in conservation planning of the Lower Keys marsh rabbit. Master's thesis, Texas A&M University.

LaFever, D. H., R. R. Lopez, R. A. Feagin, and N. J. Silvy. 2007. Predicting the impacts of future sea-level rise on an endangered lagomorph. *Environmental Management* 40:430–37.

Mabry, K. E., and G. W. Barrett. 2002. Effects of corridors on home range sizes and interpatch movements of three small mammal species. *Landscape Ecology* 17 (7): 629–36.

Morrison, M. L. 2009. *Restoring wildlife: Ecological concepts and practical applications*. Washington, DC: Island Press.

Morrison, M. L., W. M. Block, D. R. Strickland, B. A. Collier, and M. J. Peterson. 2008. *Wildlife study design*. 2nd ed. New York: Springer-Verlag.

Peters, R. H. 1991. *A critique for ecology*. Cambridge: Cambridge University Press.

Romesburg, H. C. 1981. Wildlife science: Gaining reliable knowledge. *Journal of Wildlife Management* 45:293–313.

Stamps, J. A., M. Buechner, and V. V. Krishnan. 1987. The effects of edge permeability and habitat geometry on emigration from patches of habitat. *American Naturalist* 129:533–52.

4 Habitat Management Techniques

Aldo Leopold (1933, vii) stated that "game [i.e., wildlife] can be restored by the creative use of the same tools which have heretofore destroyed it—axe, cow, plow, fire, and gun." This basic concept continues to describe the general approaches or "tools" available in modern-day wildlife habitat management. An exception to this description might include the wider variety of modern tools now available. For example, the axe that Leopold referred to might include a hydro-shear mounted on the front end of rubber-tired tractor. The gun might include the use of immuno-contraceptives to control the number of white-tailed deer in an urban area. The wildlife manager should understand all of the various options available in the practice of managing wildlife habitat.

There are many subtleties in applying these approaches that vary by geographic region. Management agencies in the various regions have often established *best management practices* (BMPs), which form a standard framework for management actions and are usually based on previous treatment outcomes, literature and other sources, and expert opinion (experience). These practices are also used for specific applications, such as to mitigate or minimize the impacts of oil and gas extraction, reduce sediment discharge in mining, and maintain stream temperatures in silvicultural operations. BMPs tend to be available as gray literature, such as agency documents or websites, and should be part of any background search when initiating a management action.

It is impractical here to provide a comprehensive review of all of these practices or approaches; instead, we review some common tools or techniques used in a very broad sense, including (1) prescribed fire, (2) mechanical treatments, (3) grazing/harvest management, (4) herbicide use, and (5) supplemental water. These are common tools likely to be applied in wildlife habitat management in various cover types (ranging from farmlands to wetlands) across broad geographic regions. The next sections discuss the function and benefit of each practice and its successful application, including planning, implementation, and monitoring and assessment. We begin by defining the various cover types found throughout the country.

Rangelands

Rangelands is a broad term that encompasses many vegetative communities. It is so broad that professionals have had difficulty defining it. For our purposes, grasslands, scrublands (dominated by shrub species), wetlands, and woodlands (less than 10% canopy closure) can all be classified as rangelands. These can be grouped according to major type of vegetation: herbaceous rangeland and shrub/brush rangeland (Anderson et al. 1976). Herbaceous rangelands are dominated by grasses and forbs but can also include those areas converted to grazing areas for domestic livestock. Much of the native grassland communities in North America have disappeared because of development and agriculture (e.g., less than 1% of tallgrass prairie remains).

Shrub/brush rangelands are typically located in drier areas than herbaceous rangelands and contain woody shrubs and trees. Like herbaceous rangelands, shrub/brush rangelands can result from human management practices, including grazing and fire suppression. The historical commonalities between these vegetative communities were a natural vegetation regime characterized by native grasses, forbs, or shrubs and strong history of natural herbivory (Anderson et al. 1976). Rangelands can also be fairly dry, which limits inclusions of wetlands to ephemeral, or seasonally dry, wetlands. Additionally, these are often early- or mid-successional communities maintained by periodic disturbance (prevented from proceeding into late-successional forests or scrublands). These disturbances can take many forms, such as wind events, floods, timber harvesting, road building, and fire. Disturbance regimes (commonly fire) have many roles in community maintenance, including impacts on succession, nutrient cycling, forage availability, palatability, nutritional content, and amount of shade.

Rangelands comprise a large portion of the vegetative communities and biodiversity in North America, for example, in Nebraska, the Dakotas, Kansas, Oklahoma, Texas, portions of the southeastern and northwestern United States, and Alaska (Anderson et al. 1976). With a high heterogeneity in micro- and macrohabitats and generally abundant cover, food, and water availability for wildlife, rangelands offer a number of niches for species to occupy. A good example is North American grasslands dominated by three very broad groups: tallgrass prairie, mixed-grass prairie, and shortgrass prairie. These grasslands serve as more than a simple residence or waypoint for herbivores. We are increasingly aware of the importance of rangelands to the maintenance of species in North America. For instance, conversion and fragmentation of more than 80% of North American grasslands

Lesser Prairie-Chicken
(*Tympanuchus pallidicinctus*)

Description: The lesser prairie-chicken is a federally threatened gallinaceous bird located in the south-central and southwestern portions of the United States. Lesser prairie-chickens inhabit some of the most endangered communities in the United States (grasslands and shrublands). Much of this habitat has been lost to agricultural activities. Additionally, they are sensitive to disturbance in these remaining core habitats as they compete with existing and emerging uses of these areas.

Lesser prairie-chicken. Watercolor and pencil image by Joyce VanDeWater from a photo by Nova J. Silvy.

Increasingly, alternative sources of energy are emerging to reduce human dependence on fossil fuels. One strategy is the use of wind turbines to produce renewable energy. However, areas with reliable wind production are often in grassland areas occupied by species such as the lesser prairie-chicken, which is known to avoid wind turbine pads and associated roads to such an extent that these areas cease to effectively support lesser prairie-chicken populations. We must also remember that turbines built in agricultural areas may reduce agricultural output and negatively impact profits, so there are costs to avoiding placement of wind turbines in prairie areas.

Scenario: Land managers wish to place nine new wind turbines in the following land area (see the figure showing farmland and prairie). In our scenario, wind turbines have a circular area of effect of approximately 1.9 centimeters (0.75 in) in diameter. Turbine areas of influence cannot overlap. Appropriately manage the lesser prairie-chicken: Determine wind turbine placement that least impacts lesser prairie-chicken populations. How will you minimize habitat fragmentation? You can choose to place the majority of the turbines in the farmland, but how would you go about convincing the farmers? If you choose to build heavily in the prairie, how would you arrange the turbines for least impact on the lesser prairie-chicken population?

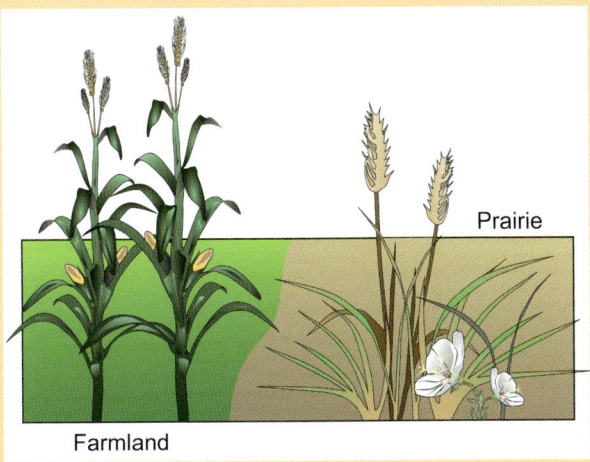

Lesser prairie-chicken habitat. Image by Joyce VanDeWater.

and scrublands over the last 200 years have coincided with a serious decline in grassland bird species (Brennan and Kuvlesky 2005).

The characteristics that made grasslands so naturally productive and capable of supporting high biodiversity (e.g., forage and grazing availability, resilience to grazing, rich soils, favorable climate regimes) also attracted anthropogenic uses such as agriculture (Heitschmidt, Vermeire, and Grings 2004). Anthropogenic rangeland uses have intensified in North America over the last few centuries. Remaining undeveloped rangelands (all types) in the United States often became associated with domestic livestock grazing. The addition of new range management strategies such as livestock grazing and fire suppression, changing climate variables such as drought, and the introduction of invasive/exotic species has certainly impacted native species' presence, abundance, and diversity (Brennan and Kuvlesky 2005).

Thus, much research has indicated the largely negative impacts on animal communities of conversion or poor management of grasslands. Carefully planned domestic livestock grazing has proven beneficial in many instances, such as using domestic goats to control plant competition in riparian restoration areas. The impact of any type of grazing (domestic or wild animals) on a rangeland is dependent on a mixture of factors, including plant structure and growth stage, water and nutrient availability, temperature, intensity of grazing (stocking rate), and species grazing patterns. Misunderstanding of any of these variables can lead to a degraded landscape with less desirable plant assemblages.

Farmlands

Farmlands are those land areas used for production of food, fiber, or fuels (Anderson et al. 1976; Heitschmidt, Vermeire, and Grings 2004). Farmlands can include the classic elements of agriculture: bush fruits, grasses and legumes, pastureland plantings, and cereal crops. They can also encompass less obvious farming components: fallow fields, failed crops, harvested croplands, horse farms, stock tanks, wetland farming such as rice fields, fish farms, irrigation structures, and farm infrastructure like roads (Anderson et al. 1976). Farmlands also include highly managed or confined operations such as orchards, vineyards, nurseries, and concentrated animal feeding operations (CAFOs). The unifying elements are generally lower human population densities and concomitant infrastructure; production or keeping of food, fiber, or fuel; and obvious farm structures such as irrigation, storage, and human residences.

The global need for food is expected to double by 2050 (Green et al. 2005) with an expected simultaneous increase in farm production (increase in efficiency and area under cultivation) and land conversion (Nelson 1992). Farmlands are a conversion of native vegetative communities into managed and simpler systems but still represent an important, if degraded, habitat for a variety of native species. These modified systems often represent important additional habitat that works in combination with protected conservation areas. Farmlands can provide food, water, and cover for many species, although not for all species and often for fewer individuals. Native species often inhabit untilled farm edges, hedgerows, and fallow fields and benefit from heterogeneity in crop rotation, field sizes, crop species, water application, and topography (Wegner and Merriam 1979; Shore et al. 2005).

Since the 1950s, farming in Europe and North America (and increasingly worldwide) has intensified with the removal of untilled areas to enlarge fields; increased planting of monocultures; more efficient removal of weeds and pests; simplification or elimination of crop rotation; and large increases in application of fertilizers, herbicides, and pesticides (Southerton 1998). Conversion of wild systems into cropland or pasture has reduced natural habitats on potential agricultural land by 50% (Green et al. 2005). Additional concerns for wildlife species' presence and abundance are pollution, diseases, further habitat loss, and emerging issues such as genetically modified organisms.

Fortunately, agriculture and biodiversity need not be entirely mutually exclusive. Farming often benefits from the presence of wild species (e.g., diverse bird assemblages that help control invertebrate agricultural pests; Kellerman et al. 2008). These issues have precipitated a vigorous debate about land sparing and wildlife-friendly farming practices (Fischer et al. 2008). Land sparing involves intensely farming (e.g., monocultures, intensive herbicide and pesticide use) certain plots of land and avoiding farming adjacent plots. The goal is to maximize production on the agricultural plots as intact vegetative communities are preserved on other plots. Species sensitive to agricultural operations could benefit since relatively undisturbed areas are maintained. On the downside, no land management strategy exists in a vacuum, and the actions on intensively farmed agricultural areas can cross to other lands (e.g., water pollution). Rather than sacrifice some plots and spare others, wildlife-friendly farming lowers farming intensity by combining patches of native vegetation and crop heterogeneity on agricultural lands (Fischer et al. 2008). This might include planting more varieties of crops rather than monocultures; using diverse rotation strategies; leaving natural vegetation areas; and decreasing use of herbicides, pesticides, and fertilizers. This strategy would likely help those species adapt to low-level agricultural operations and allow species movement through habitat connectivity.

Forests

Forests are technically defined as areas that have a tree-crown areal density (amount of area covered by tree canopy) equal to or greater than 10%, but they are far more than a simple measure of crown-closure percentage (Anderson et al. 1976). They are providers of ecosystem services such as flood control, carbon sequestration, and timber and are critical for biodiversity. North American forests are managed for a variety of purposes, including timber extraction, ecosystem conservation, recreation, municipal needs like water provision, and grazing. There are approximately 304 million hectares (751 million ac) of forest in the United States comprising roughly 33% of the land area and about 800 species of trees (Smith et al. 2002). This is a major decline from estimated forested area in 1630 (405 billion ha, or >1 trillion ac) with much of that conversion occurring in 1850–1900 but continuing throughout the twentieth and twenty-first centuries.

Forests can be placed into three broad categories: (1) deciduous, (2) evergreen, and (3) mixed forests (Anderson et al. 1976). *Deciduous forests* are mainly composed of trees that lose their leaves at the onset of the cold or dry seasons (water oak [*Quercus nigra*] and aspen [*Populus tremuloides*]). *Evergreen forests* are dominated by trees that remain green through the entire year (loblolly pine [*Pinus taeda*] and hemlock [*Tsuga canadensis*]). *Mixed forests* have both deciduous and evergreen trees in roughly equal amounts (greater than 33% of both forest types). Forests themselves are characterized by several important attributes: (1) composition, (2) function, and (3) structure. Composition describes which species are present and in what numbers. The forest functions are the processes carried out by the forest ecosystem. Forest structure describes spatial arrangement and characteristics of landscape components like trees, logs, and snags (standing, dead, or dying trees; Franklin et al. 2002).

Forests must often support timber extraction, maintain functioning ecosystems, protect biodiversity, and allow human recreation and development, sometimes all at once. Most often, these uses require trade-offs (DeFries, Foley, and Asner 2004) between resource use and ecosystem function (Corbett, Lynch, and Sopper 1978; Crocker-Bedford 1990).

Disturbance regimes are critical to proper functioning of forests but are often seriously altered by human influence such as fire suppression and flood control. The suppression of disturbance can lead to changes in plant regimes and a simplified ecosystem (e.g., fewer canopy openings), which then impact other species assemblages and ecosystem function. Today, many forests require complex management strategies that must integrate with human needs. For instance, in areas where prescribed fire is untenable, mechanical removal can mimic natural disturbance regimes (Pike, Webb, and Shine 2011). The important message is that multiuse forests are increasingly the norm, but complex uses lead to difficult social and ecological problems that require a well-stocked forest management toolbox.

Wetlands

Wetlands have water tables near, at, or above the surface (Anderson et al. 1976). They come in all shapes and sizes (e.g., marshes, salt marshes, bogs, ponds, littoral areas), but all have a central component of water at least some of the time. Hydrology is foundational in wetlands in that it impacts soil characteristics, nutrient availability, and salinity, which all influence biota (the collection of organisms at a given time; Mitsch and Gosselink 2000). Biota, in turn, also greatly influences the ecosystem in a variety of ways (e.g., beaver dams, leaf detritus impacts on soil nutrients).

Wetlands generally fall into one of two very general categories: (1) forested and (2) nonforested. *Forested wetlands* have a large woody vegetation component and include familiar variations such as cypress swamps and seasonally flooded forests (Anderson et al. 1976). *Nonforested wetlands* lack a dominant woody vegetation component and comprise primarily herbaceous vegetation or are unvegetated. These include some of our stereotypical visions of wetlands such as freshwater meadows and salt marshes. Some wetlands do not conform easily to these general categories. Seasonal wetlands lack water at certain times of the year, and some alluvial flats, or playas, may have no vegetation at all. Some agricultural lands, such as cranberry bogs, rice fields, and peat bogs, are technically wetlands. Seasonally inundated floodplains (e.g., Nile River floodplain) and regularly flooded irrigated areas are considered wetlands as well (Anderson et al. 1976).

Wetlands are generally very productive systems. They can have high primary productivity (energy storage in photosynthetic species) and concomitant high numbers of dependent plants and animals. With high energy availability, abundant water, and plenty of cover, wetlands are important spawning and feeding areas, nurseries, and migratory destinations or rest areas. They directly support many wildlife species and indirectly support other ecosystems through ecosystem services such as water purification. Additionally, the high productivity of wetlands combined with strong selection pressures has resulted in the presence of many endemic species (those unique to a defined location; Gibbs 1993). Thus, the degradation or loss of wetlands has large impacts on conservation and wildlife management, especially since there are many currently threatened and endangered species native to wetlands (Gibbs 2000). These native species often perish as their wetland habitats are altered. Consider how difficult it is for many wetland species to move as habitat is damaged or destroyed, such as the Santa Cruz long-toed salamander (*Ambystoma macrodactylum croceum*). Salamanders are unlikely to move far overland to find another wetland source.

The conservation of wetlands began decades ago with laws such as the Migratory Bird Conservation Act (1929) and the Migratory Bird Hunting and Conservation Stamp Act (Duck Stamp Act, 1934) that protected migratory bird habitats. Hunters, conservationists, recreationalists, and legislators worked together to protect many ecosystems, including wetlands. The Duck Stamp Act has protected more than 1.8 million hectares (4.5 million ac) of wetlands for waterfowl (Mitsch and Gosselink 2000). The 1960s and 1970s brought a variety of important conservation laws relevant to wetlands, including the Endangered Species Act (1973) and the Federal Water Pollution Control Act (1972). However, we have lost at least 50% of the wetlands worldwide and 50% of the original wetlands in the United States (Mitsch and Gosselink 2000). Current wetlands are threatened with pollution, invasive exotics, hydrologic changes, and urbanization.

Urban Areas

Urban areas are dominated by human-manufactured structures, such as those in residential, commercial and services, industrial, transportation, communications, utilities, and mixed urban areas (Anderson et al. 1976). Residential areas are places where high numbers of humans choose to live. Although human population density differs between residential areas, such areas tend to consist of regular organization of structures. Commercial and service sectors often commingle or exist adjacent to residential areas to provide goods and services for the human population. These might include shopping and strip malls, restaurants, educational centers, and waste disposal operations. Industrial areas are a broad assortment of land use types with particular emphasis on production and manufacturing. These often include factories, refineries, power-generation plants, mills, and rail yards. They might have open stockpiles of raw materials or waste products. Transportation, communications, and utilities can overlap significantly with industrial areas (e.g., power plants) but are often unique structures with large footprints (e.g., highways). Mixed urban areas are a complex interweaving of multiple land uses such as residential areas built in industrial sectors or combined industrial-commercial complexes.

The common thread in all of these types of urban areas is a conversion or fragmentation of native habitat to build structures. Approximately 80% of the US population lives in expanding urban or suburban areas (McKinney 2002). This urbanization dramatically impacts ecosystem function and services. McKinney posits that the edges of developed areas have natural remnant vegetation, but as we progress toward the center, this is replaced with ruderal (early colonizers of disturbed areas) vegetation consisting of damaged, unmanaged green space. Further in, we run into managed vegetation like backyards and other maintained green spaces. Finally, we encounter the city center dominated by buildings and sealed surfaces (e.g., concrete roads). As we should expect, this gradient corresponds to declining biodiversity from the edge of developed areas to the center.

Wildlife is acutely impacted by urbanization, though the degree is dependent on a number of factors (human-built structure density, socioeconomic factors, individual species ecology, conspecifics, etc.). Impacts on wildlife can include factors such as reproductive success and disease occurrence (Boal and Mannan 1999).

Even with its variable impacts on plant and animal species, urbanization has undoubtedly played a major role in biodiversity declines. There are several major reasons for this: urban areas often permanently replace native habitat, species in urban areas are usually weedy (rapid colonizers, quick reproduction) and nonnative, and urban areas often provide worse habitat for species than the native landscape (McKinney 2002). Fortunately, urban and suburban area biodiversity can improve through active planning and management. Human communities can choose to manage for and protect extant native species, for example, by promoting the growth of native plant species in residential and communities areas.

Habitat Management Techniques

Various tools can be used to manage habitat, such as prescribed fire, mechanical treatments, grazing, harvest management, herbicide applications, and provision of supplemental and replacement water.

Prescribed Fire

Prescribed fire (Aldo Leopold's "fire") is perhaps one of the oldest and most economical approaches to improve, maintain, or restore various types of wildlife habitats, particularly those that are fire-dependent, such as longleaf pine savannah. Prescribed fire, also known as *controlled fire* or *prescribed burning*, is a deliberate or intentional use of fire under specified and controlled conditions to manipulate vegetation for a given set of management objectives. Through this process, prescribed fire is applied in a safe manner to achieve these objectives, reduce the risk of wildfire, and ultimately improve the overall condition of wildlife habitat. Some example objectives might include stimulating new plant growth, removing dead vegetation, or creating early-successional conditions. The general public is traditionally fearful of fire because it is viewed as being extremely destructive; as a result, prescribed burning is often underutilized. Furthermore, the application of prescribed fire is becoming more and more difficult due to increased urbanization and air-quality restrictions in urban-rural landscapes. Despite these management challenges, the use of fire has many benefits in habitat management and continues to be a cost-effective approach to consider.

Historically, the use of fire played a significant role in many cultural practices (Hankins 2005) and dates back approximately 1.0–1.5 million years. In fact, the "purposeful use of burning is often ranked among the attributes which elevated humans to dominance of the world ecosystem—the refined capability to reason, ability to make and use tools, and upright posture" (Scifres and Hamilton 1993, 37). In addition, many early human economies were dependent on the use of fire as a means to clear lands for agricultural purposes (Sauer 1950); thus, the application of fire in maintaining or shaping natural systems was long utilized in human history to include the management of wildlife habitats.

Function and Benefits

There are many benefits in use of prescribed fire in wildlife habitat management, including (1) controlling the vegetative structure, (2) improving the palatability and nutritional value of wildlife plants, (3) controlling diseases and pests, and (4) reducing the risk of wildfires.

Varying the frequency and timing of prescribed fire will often yield different results when applied (Sargent and Carter 1999). For example, a summer burn in a forested area can result in different understory plant communities than a winter burn because of differing seed germination requirements.

A major benefit of prescribed fire is the manipulation of vegetative structure, such as controlling the number of saplings or reducing the amount of woody midstory/understory vegetation. Prescribed fire is a form of disturbance that can serve to "restart" succession in many types of vegetation and is a valuable approach in habitat management in many ecosystems. In fire-dependent forested areas, for example, the manipulation of vegetation through prescribed fire serves to control understory growth, reduce leaf litter, and recondition the soil through the release of nutrients. It also shapes many fire-dependent ecosystems. In the southeastern United States the use of fire in longleaf pine forest savannas is critical to their maintenance. For instance, hunters, bird-watchers, and land managers can access areas free of thick underbrush. It is much easier to hunt, search for birds, or mark a stand of timber in an area recently burned. So improved vegetative structure not only is beneficial from a wildlife perspective but also from a land management perspective where improved access can achieve management goals much more efficiently.

Another benefit of prescribed fire is improvements in the palatability of forage and browse plants. The quality and quantity of grasses, forbs, legumes, and other herbaceous plants can be improved and increased through its use. These plants can in turn provide cover and food for numerous game and nongame species. Many shrubs, grasses, and legumes produce more fruit in recently burned areas than in non-burned areas. Furthermore, in general, wildlife species prefer recent vegetative growth or browse. It is difficult for some species to consume and digest older grasses, for example, than new, succulent plant shoots resulting from a recent fire. Likewise, regrowth following a burn often increases the abundance of arthropods that serve as food for songbirds and their nestlings as well as other insectivorous animals.

Destructive pathogens such as fungi, bacteria, and other disease-causing organisms as well as common pests like chiggers and ticks can be controlled and managed through the use of prescribed fire. Diseases affecting plant foliage can be burned to eliminate pathogens and subsequent transmission. Invasive plants also threaten both the survival of native plants and the native wildlife populations dependent on these plant communities. Prescribed fire can be used as an effective control of invasive plants. In rangelands, for example, prescribed fire can be used to control invasive vegetation such as prickly pear (*Opuntia* spp.). Control of invasive aquatic plants like giant salvinia (*Salvinia molesta*) can restore watershed functions and improve water quality and quantity.

The removal of *thatch*, or vegetation accumulation, can reduce hazardous fuel loads and decrease the risk of wildfires. Common sense tells us that it is hard to burn in an area where the necessary fuels are no longer present. Thus, the use of prescribed fire is an excellent wildfire prevention mechanism in maintaining fire-dependent plant communities. The increase in wildfires across the United States typically is attributed to increased fuel loads due to a lack of regular, prescribed fires. This pattern of wildfire is exacerbated in fire-dependent plant communities. The risk to humans living in these fire-dependent landscapes without the regular use of prescribed fire will likely continue to be a safety concern, particularly with increased urbanization (often referred to as the *urban-wildland interface*) that limits the safe application of fire. Furthermore, the continued variability in climate also will lead to prolonged drought conditions and increase the risk of wildfires.

Application

About 90% of prescribed fire use involves activities that occur *before* striking a match. Most of these activities involve necessary steps for the safe execution of prescribed fire centered on preparation and preparedness. We begin by revisiting the idea of goals and objectives and some basic concepts regarding fire theory.

Before initiating a prescribed burn, land managers must develop a fire management plan that includes goals and objectives. The importance of clear and executable objectives and goals cannot be overemphasized; they are necessary to evaluate the success of a prescribed fire in producing the desired outcomes. If the objective is, for example, to suppress or control woody plants, invasive plants, or other noxious plants, the fire plan must incorporate the necessary actions to prepare for, execute, and complete the requirements to meet the plan's goals and objectives. Additionally, the plan needs to identify long-term environmental goals, weather conditions projected for the date of the burn, list of equipment and supplies, and the necessary personnel required to safely conduct a prescribed burn (Scifres and Hamilton 1993). Comprehensive prescribed fire plans will minimize risks (probability of an unwanted event) and increase desired results; omitting or skipping steps can lead to catastrophic outcomes and cause severe damage to neighboring lands, livestock, and wildlife.

The manager must understand the basic elements of fire combustion, also known as the *fire triangle* (fuel, heat, and oxygen; fig. 4.1), in the safe execution of a prescribed fire. The removal of one or more of these elements prevents or extinguishes a fire; thus, the use of prescribed fire is under the control of the manager, who determines the presence of these elements. In other words, a fire is controlled through the land manager's preparedness in removing one or more of these elements during the process of applying fire to an area.

This flexibility obviously requires preparation and training on the part of the manager, for example, the use of *firebreaks*, areas where the *fuel* component of the fire triangle is physically removed to prevent the continued spread of fire beyond the area of interest. Obviously, establishing firebreaks would be an important first step executed by the manager *before* lighting a match. Some examples of firebreaks include a disked line, a roadway, or a water body. Burning and extinguishing fuel along the perimeter of an area targeted for a burn (often referred to as *black lining*) also can create a firebreak through the removal of fuel, using fire rather than mechanical or other physical measures. Often, the two techniques can be combined through the reinforcement of firebreaks by black lining an existing firebreak, particularly in an area where escape of fire is possible (Scifres and Hamilton 1993). We discuss the application of firebreaks in more detail later; however, the point we make here is that understanding the fire triangle can help safely execute a prescribed fire in habitat management.

Planning

The former heavyweight champion boxer and philosopher Mike Tyson once said, "Everybody has a plan until they get punched in the face." With prescribed fire, you need a strong plan to reduce the risk of getting punched in the face! Some pre-fire considerations the manager

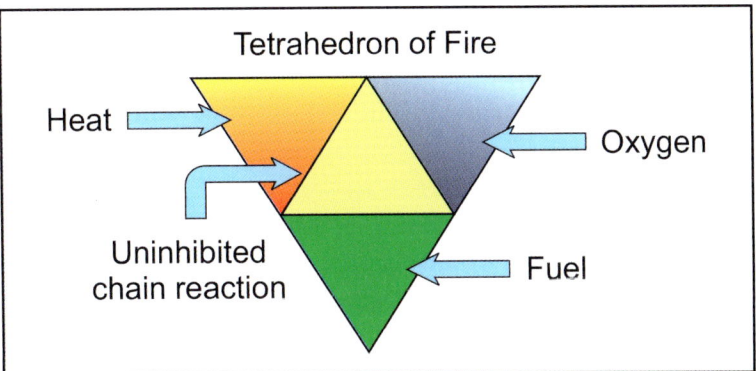

Figure 4.1. Tetrahedron of fire. From https://en.wikipedia.org/wiki/Fire_triangle#cite_note-3.

should take into account include timing effects, environmental factors, and associated costs.

The timing in the application of prescribed fire is closely tied to the management goal and objectives for the area of interest. Most prescribed fires are conducted during the fall and early spring months, primarily when fuel conditions are ideal and when weather conditions also provide a safe environment to conduct a prescribed fire. Most prescribed fires are conducted during this time period, or "window"; however, it is important to note that historically fires were ignited during the late spring and summer months when storms and associated lightning strikes were most prevalent. The reason most prescribed fires are conducted in the winter and early spring is primarily due to safety. These *cool-season burns* are less intense so are typically favored by land managers and the general public, who are concerned with liability issues such as escape of the fire.

Summer burns, or "hot" fires, are used in some instances such as during site preparation of a recently cleared forest area. Summer burns reduce heavier fuel loads like slash and debris that are less likely to burn during cool-season burns. Nonetheless, prescribed burning needs to be conducted when target vegetation will likely burn most favorably (Hanselka 2006). Understanding the likely vegetative response resulting from the fire will aid in determining the appropriate time of year to conduct the burn. This also means that managers must understand wildlife needs and likely wildlife response to altered vegetation. For instance, they need to plan the fire to ensure adequate cover and food supplies during critical post-reproduction months for grassland bird nesting seasons.

The first major consideration in executing a prescribed fire is weather, "the state of the atmosphere at a given time and place, with respect to variables such as temperature, moisture, wind velocity, and barometric pressure" (Federal Aviation Administration 2008, 11–1). Ideal burning conditions are primarily dependent on optimal weather. Thus, a major challenge is the difficulty in forecasting optimal weather conditions, particularly as weather can easily change from one moment to the next. For these reasons, fire crew members need to monitor weather conditions prior to and during a prescribed fire, especially wind direction and velocity. Three important measures of weather are relative humidity, temperature, and wind speed/direction.

Relative humidity is the ratio of partial pressure of water vapor in the gaseous mixture of air and water vapor to the saturated vapor pressure of water. In other words, it is the amount of water vapor that exists in a gaseous mixture of air and water vapor. The ideal relative humidity for prescribed burns is between 50% and 70%. Relative humidity below 50% results in intense or "hot" fires primarily due to dryness of fuel such as grasses and vegetative debris. Relative humidity above 70% is likely to make the application of prescribed fire difficult or lessen the fire's ability to "carry" through the area in a uniform way (Sargent and Carter 1999).

Temperature, which is closely related to relative humidity, also plays an important role. When the temperature is below 0°C (32°F), for example, grass will rarely burn; when it is above 27°C (80°F), that same grassy area will burn extremely hot and can easily become a dangerous situation. In general, the optimal temperature for prescribed fires is between 4.4°C and 15.5°C (40°F and 60°F). In addition, the direction and speed of the wind should be consistently monitored before and during the burn because any wind variance can significantly alter fire behavior. The optimal wind direction and speed should be steady and between 4.8 and 12.9 kilometers per hour (3 and 8 mph; Sargent and Carter 1999). Wind gusts can cause vegetation to burn too quickly and make fire control difficult.

Another important consideration in the application of prescribed fire is the available fuels for fire consumption. *Fuel* is defined as any combustible material, such as grasses and woody vegetation. Fuels are characterized by factors such as vegetation compactness, horizontal continuity, vertical arrangement, chemical content, and moisture level. Most fire managers characterize available fuels using the National Fire Danger Rating System (NFDRS), a fire standard used across the country. Using the NFDRS classification, fire managers categorize fuels according to how they respond to changes in moisture, referred to as *time lag*. There are four fuel categories:

- 1-hour fuels: <0.6 centimeter (<0.25 in) in diameter
- 10-hour fuels: 0.6–2.5 centimeters (0.25–1.0 in) in diameter
- 100-hour fuels: 2.5–7.6 centimeters (1–3 in) in diameter
- 1,000-hour fuels: 7.6–20.3 centimeters (3–8 in) in diameter

Examples of 1-hour fuels include grasses, leaves, mulch, and litter. From a prescribed fire perspective, 1-hour fuels are generally the focus, as fuels in this category serve to "carry" most prescribed burns. An adequate amount of 1-hour fuels is needed to ensure that the prescribed fire is uniform and complete. By definition, the moisture level in 1-hour fuels can change within an hour's time lag due to environmental factors like temperature, rain, humidity, and shade. Conversely, larger fuels such as large trees or brush piles require 100+ hours to respond to changes in environmental factors. From a prescribed fire perspective, less than 10-hour fuels are generally measured and monitored prior to burn execution.

Related to fuel type is *fuel loading*, the dry weight of grasses and leaf litter in a given burn unit, usually expressed in pounds per acre. Similar to estimating forage production, fuel load estimates can be obtained visually by experienced land managers or through vegetation measurement (e.g., measurement of biomass using clipping, measurement of visual obstruction using a range pole). Both the types of fuels and amounts are important considerations in prescribed fire. Understanding the relationship between fuel structure and ability to control fuel moisture allows the fire manager to control a fire. This control is defined within a given prescription, an outline of environmental parameters where a prescribed burn can be applied both safely and effectively. A prescription outlines, for example, the preferred relative humidity, fuel loads, and wind speed/direction needed to execute the burn. More on this process is discussed in the implementation section.

As we have emphasized thus far, the majority of the control in a prescribed fire actually occurs prior to ignition. In addition to the timing issues and other environmental considerations, conducting a prescribed fire should include training, permitting, and managing smoke. In terms of training, fire crew members should be knowledgeable of the fire basics. Various agencies like the Natural Resources Conservation Service (NRCS), cooperative extension agencies, state forestry commissions, and state wildlife agencies can assist through training courses and workshops. In some states, fire crew members need to be "burn-certified" to legally conduct a prescribed fire. Course work and field experience are typical requirements associated with being certified in many states. In addition to certification, obtaining the necessary fire permits and liability insurance in the unlikely chance of a fire escape may be required to legally implement prescribed fire. Liability insurance is a significant challenge to implementing prescribed fire, particularly on private lands. Typically, farm insurance plans (e.g., Farm Bureau insurance) include provisions for liability associated with controlled burning; however, the user must read exclusions often tied to these insurance policies.

Proper notification (e.g., local fire department, sheriff's department, neighbors) of the plan to burn is strongly recommended and in some cases mandatory. Once the fire is completed, renotification of these primary contacts needs to be conducted to inform them the fire has been safely contained. Finally, a growing issue related to prescribed fire use is the management of smoke, which should consider wind direction and areas likely to be impacted by low visibility and risks to human health. Burning away from roadways or neighboring communities is an important consideration for land managers primarily due to associated liability issues. An accident on a state highway due to low visibility caused by a prescribed fire makes the manager liable (Sargent and Carter 1999). A checklist can help ensure that proper preparation and training are in place prior to executing a prescribed fire (table 4.1; Scifres and Hamilton 1993).

Costs

Prescribed fire is an economical and cost-effective practice in the manager's toolbox for managing wildlife habitat. The cost per acre varies regionally and is associated with several factors, including the size of the target area (larger tracts are cheaper to burn on a per-acre basis) and the degree of infrastructure that needs protection (homes, power lines, etc.) via firebreaks or additional crew members. Some common costs to consider include labor, materials (fuel for drip torches, water, etc.), and firefighting equipment (flappers, fire rakes, fire pump, etc.; table 4.2). Liability insurance is probably the most significant cost associated with prescribed fire. While even the best prescribed fire plans can successfully identify proper equipment, materials, pre-burn preparations, and labor needs, the possibility always exists for fire to escape and cause unintentional damage. In some states like Texas, liability insurance ($1 million in coverage) is actually required for someone to be burn-certified. Despite all these costs, prescribed fire continues to be a cost-effective approach (around $5/ac) compared to other land management treatments like bulldozing, cutting, or chemicals (greater than $30/ac; Sargent and Carter 1999).

Implementation

Once all the necessary safety precautions are in place, the execution of a prescribed fire can begin. The land manager should have basic knowledge in (1) fire-ignition techniques, (2) types of fires, and (3) fire suppression and containment. Before beginning fire ignition, the manager must reevaluate the weather at the field site to make sure the fire will occur within prescription. Notification of the planned fire must be made to the proper authorities to keep everyone informed. All equipment should be checked, including communication equipment for field personnel, such as two-way radios, and suppression equipment to make sure the fire pump/tank is working. Once all is in place, the actual ignition can begin.

Table 4.1. Summary checklist for prescribed burning

Pre-burn rationale

    Identify and target pastures

    Purpose (reclamation, maintenance, etc.)

    Timing (cool or warm season)

    Consideration of vulnerable areas (erosion, wildlife, etc.)

    Legal responsibilities (liability)

Planning the burn (planning is conducted 6 months to a year in advance)

    Consider alternative feeding locations for livestock, wildlife, etc.

    Identify potential burn dates at least 12 months in advance

    Schedule burning dates 6–8 months prior to the burn

    Identify budget cost

    Identify and set burn dates

    Identify number of required trained individuals

    Complete a comprehensive prescribed fire plan

Post-burn considerations

    Patrol burned areas to ensure fire does not reignite

    Inspect fences, posts, electric poles, etc.

    Monitor smoldering piles

    Check for livestock and wildlife access

    Observe vegetation changes

    Project grazing periods and monitor change observations

    Identify and evaluate when burned areas will reach optimal conditions to restock livestock (calculate proper stocking rates)

Prescribed fire action plan

    Identification of burn site (must be completed 6–12 months in advance)

    Area location

    Area size (in acres)

    Proposed date(s) (12 months in advance)

    Set date(s) (6–8 months in advance)

    Outline burn area perimeter; also sketch area with as much detail as possible, i.e., include topographical information when possible, indicate firebreaks and escape routes

Identify and list prescribed fire objectives

Fuel reduction	Woody plant control
Weed control	Forage improvement
Increased wildlife habitat	Disease control
Improved wildlife diversity	Reduce understory growth
Other	

Fuel load accumulation practices

    Grazing deferment (approximate date when deferment began)

    Chemical treatment (type and rate, date applied)

    Mechanical treatment (type and dates conducted)

Site characteristics

    Percent of grasses, forbs, and legume cover

Percent of low-volatile plant cover (e.g., honey locust)

Percent of high-volatile plant cover (e.g., eastern redcedar)

Percent of leaf matter and other debris

Identify slope and terrain of the site

Level terrain	Level to rolling terrain
Moderately steep	Steep

Identify and list all potential hazards near the perimeter of burn site (electrical lines and poles, gas wells, barns, etc.)

Identify and list bodies of water, points of access, location of firebreaks, etc.

Acceptable weather and moisture conditions

Wind velocity	Wind direction
Relative humidity	Temperature
Soil moisture	Plant cover moisture
Other	

Apply for permits

Complete within one month before prescribed burning and notify the proper authorities and neighbors. Also, secure equipment, fire crew members, and emergency assistance information.

Obtain contact information, i.e., name of agencies, address, phone number, website, contact personnel's e-mail address, and date permit was issued.

Notify state and local agencies	Address, telephone numbers, websites, e-mail addresses	Contact personnel
Fire department		
State police		
Local police		
Sheriff's department		
Neighbors		
Other agencies		

Equipment requirements	Item description	Model number	Source
Burning equipment			
Safety equipment			
Other			

Crew requirements

Number of Crew Members

Crew member and contact information	Address, telephone numbers, websites, e-mail addresses	Contact personnel
Fire boss		
Fire members		
Emergency and first-responder assistance	Address, telephone numbers, websites, e-mail	Contact personnel

Table 4.1. (*continued*)

Firefighter

Red Cross

Emergency first responders

Other

Firebreak construction (complete 2–3 weeks in advance)

Plowed firebreak	Disked firebreak
Mow	Natural firebreaks (lakes, rivers, creeks, etc.)
Other firebreak construction	

Prescribed fire preparation activities (complete day before or day of prescribed fire)

National weather and/or local weather service contact information

Pre-burn discussion with fire crew members

Prescribed fire objectives and goals

Target burn area

Chain-of-command communication plan

Tour of targeted burn area

Fire-ignition patterns and plan of action

Smoke-management plan

Firebreak inspection

Equipment inspection and use demonstration

Outline and discuss hazards

Worst-case scenarios and action plans

Identify and locate assistance resources

Inspect crew members for proper clothing

Crew assignments and responsibilities

Ignition

Suppression

Direct traffic

Fire data recording

Date and time of prescribed fire ignition

Weather conditions

Beginning of burn	End of Burn
Wind speed	Wind speed
Wind direction	Wind direction
Relative humidity	Relative humidity
Air temperature	Air temperature

Soil moisture content: saturated moist   dry

Identify smoke-monitoring procedures and action plan to minimize and/or control smoke during the burn

Results and description of fire-behavior test

Type of ignition patterns used to start fire

Flank fire	Location	Rationale
Head fire	Location	Rationale
Backfire	Location	Rationale

Strip-head fire	Location	Rationale

Post-burn activities
 Time fire was extinguished
 Equipment collection/rental equipment return
 Burn area perimeter inspection
 Burn completion notification to local and other authorities
 General remarks and observations (weather changes, equipment problems, etc.)

Drip torches are probably the most common method for igniting a fire. They are hand-carried devices that pour out a small stream of burning fuel onto the ground safely without igniting the device (fig. 4.2). The fuel mixture used in a drip torch is typically half diesel and half gasoline. Drip-torch-ignited fires are typically low intensity, allowing the applicator to light the identified path or section by walking in the direction necessary to accomplish the objectives of the fire. Another form of fire ignition is the sphere dispenser machines ("Ping-Pong" balls) mounted to an aircraft and dropped from the air. These *delayed aerial ignition devices* (DAIDs) are polystyrene balls 3.18 centimeters (1.25 in) in diameter that contain potassium permanganate. The balls are fed into a dispenser where they are injected with a water-glycol solution and then dropped through a chute. The chemicals react thermally and ignite in 25–30 seconds. Or flame-throwing devices can be mounted onto a four-wheeler or helicopter, the latter known as a helitorch. These ignition devices are mounted to disperse ignited lumps of gelled gasoline. Regardless of the ignition process, it is important to use caution and ignite fires with precision in the direction necessary to accomplish prescribed fire goals.

Various types of prescribed fires are defined based on the wind direction and associated fire behavior. The speed and direction of wind affect fire behavior by the addition of oxygen (one of three elements of the fire triangle) and direction of the fire's movement. *Head fires* are set with the wind and are fast moving and potentially dangerous due to rapid movement, flame height, and fire intensity. If properly applied, head fires can be extremely economical due to the quick application and termination of a prescribed burn; however, head fires should be applied by experienced applicators. Head fires can also be effective in the rapid reduction of fuels due to the higher associated fire temperature or intensity.

Table 4.2. Prescribed fire cost checklist

Fuel	**Grazing loss cost (lease rate)**
Diesel	**Labor cost for actual burn**
Gasoline	Number of personnel × daily rate
Firebreak preparation	**Other labor-related expenses**
Mowing	Travel expenses
Raking	Per diem
Dozing	Overnight accommodations
Labor for firebreak preparation	**Protective gear**
Purchase tools	Heat-resistant clothing, hats, gloves
(5-year life)	Eye protection
Drip torches	**Drinking water**
Fire rakes	**First-aid kit**
Backpack sprayers	**Belt weather kit (one per burn)**
Rental equipment cost	**Liability insurance**
4-wheeler	**Filing fee (if applicable)**
4-wheeler sprayer	**Contracted services**
Cattle sprayer	**Other miscellaneous expenses**
Bush hog	**Total cost**
Tractor	**Total acres burned**
Fire truck	**Total cost/acre**
Fire igniter	

Backfires are set at a 90° angle to the wind, resulting in the backfire burning directly into the wind. Backfires are considered the safest type of fire and are commonly used to secure both existing firebreaks by increasing the amount of black line or burned areas adjacent to a firebreak and to observe initial fire behavior. Firebreaks are areas where the fuel component of the fire triangle is physically removed to prevent the continued spread of fire beyond the area of interest and extremely important in the containment of a prescribed fire. Backfires increase the width of firebreaks through the removal of fuel through fire combustion. Once the target area is backfired, the chance of a fire "jumping" a firebreak is minimized and other types of fires like flank fires or head fires can be used.

Flank fires are often used with low fuel loads and ignited by individuals walking into the wind. Flank fires have some of the characteristics of head fires but typically extinguish themselves quickly when they burn into the firebreak. Often, parallel strips of head/flank fires are lit to obtain the benefits of more intense fires without the associated danger. *Spot fires* are sometimes used in igniting a prescribed fire. They are ignited at equidistant locations throughout the target area and gradually expand and join each other. Spot fires are typically used in areas where good firebreaks are established and large areas are projected to be burned. For example, a burn of 405+ hectares (1,000+ ac) would be difficult to ignite with drip torches. Helicopters using DAIDs can quickly ignite an area via aerial ignition using spot fires.

All fire types are known to have both positive and negative attributes, and the type of fire used in ignition greatly depends on weather conditions, fuel load, size of the area, and expertise of individuals assisting with the prescribed fire (fig. 4.3). Regardless of type of ignition, fire crew members must monitor the progress of the burn, patrol and observe fire behavior, and, if necessary, take actions to ensure fire safety.

As we have emphasized, the first line of defense in conducting a prescribed fire is having adequate firebreaks surrounding the proposed burn site. In the event that fire suppression is needed (e.g., the escape of fire beyond the target area), tools such as flappers, rakes, shovels,

Driptorch, burning fuel, matches, fire ignitor	Water equipment such as fire truck, tractor mounted sprayer, backpack sprayer, etc.
Hand tools such as rakes, shovels, etc.	Protective gear such as leather boots, gloves, goggles, glasses, etc.
Cellular phones, handheld radios, etc.	
First Aid kit, bottled water	Tractor with disc or dozer

Figure 4.2. Commonly used prescribed fire equipment. Image by Joyce VanDeWater.

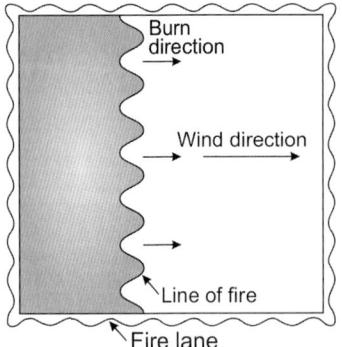

Head fire

Head fires are set with the wind direction and should only be used by experienced professionals under ideal fuel conditions. Often set after rain, head fires are the most economical and the most dangerous type of prescribed fire. Head fires burn quickly, have a taller flame, and can kill even large pines if used improperly. If used properly, they are very effective at maintaining early successional wildlife habitat.

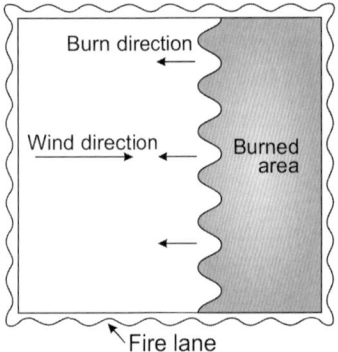

Backfire

A backfire is set at a 90-degree angle to the wind direction so the fire burns directly against the wind. This is one of the safest methods of prescribed burning and is recommended for beginning wildlife managers or where fire hazards exist, such as adjacent lands with high danger fuels. Wind speed should be no more than 6-10 mph. At night, backfires normally move at about 1 chain (66 feet) per hour. If the wind velocity is 20 miles/hour, the fire will back twice as fast (132 feet/hour).

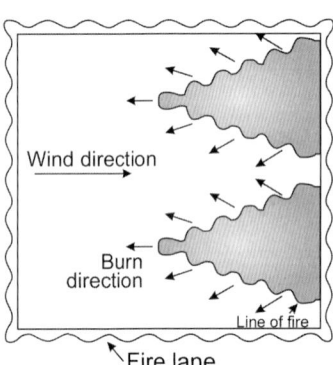

Flank fire

Flank fires are often used when the fuel is relatively light. These fires are set by an individual or individuals walking into the wind and are relatively safe.

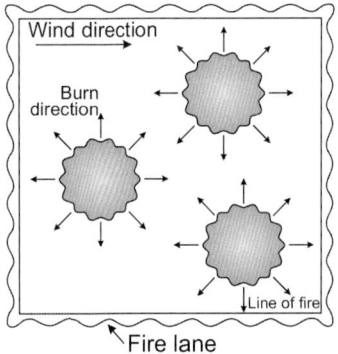

Spot fire

Ideally, spot fires are set at equidistant locations throughout the forest. These fires gradually expand until they join.

Figure 4.3. Wind and burn directions of head fire, backfire, flank fire, and spot fire. From Strickland, Edwards, and Hamrick (2016).

and pulaskis (modified fire axes) can be used in direct or indirect fire suppression. A *fire flapper* or *swatter* is a large rubber mat attached to a wooden handle that smothers a fire through the removal of oxygen. Other suppression tools such as fire rakes and shovels also rely on the removal of fuel and/or oxygen. Other commonly used fire-suppression tools spray water or other fire retardants directly on a fire. These *fire pumps* can range from units on large fire trucks to those on an ATV or tractor. Often these fire pumps use gasoline-driven water pumps to deliver water under pressure via a hose or other similar device or nozzle. Specialized backpack sprayers, often referred to as bladder pumps, and common household garden sprayers are handheld fire pumps. Each of these tools is useful in safely implementing a prescribed fire and being able to combat a fire in the event of escape. Of course, the equipment should be properly maintained and tested prior to fire ignition.

Following the completion of the prescribed fire, the fire manager should take the following important steps. First, the proper authorities should be contacted to inform them that the prescribed fire is no longer active. Second, the fire crew should go through the process known as "mop up." *Mop up* occurs after a fire, or any part of a fire, is controlled and is the process that makes a fire safe by extinguishing or removing burning and hazardous material. The fire crew should never leave the site until proper mop up has occurred. The process of mopping up can include extinguishing all smoldering material along the burn perimeter, ensuring logs/debris cannot roll or fall across the fire line, or searching for underground burning roots. These potential fire hazards should be properly extinguished using either water or soil to prevent the recurrence of a fire. Once a fire is controlled and mopped up, the fire boss can then declare the fire to be complete.

Monitoring and Assessment

An important last step following the execution of a prescribed fire is monitoring and assessment. A well-designed monitoring and assessment protocol can help with management plan revisions if for some reason the management goals were not achieved. Did the quality of wildlife browse improve? Did the amount of fuel loads decrease, resulting in a lower risk of wildfire? These questions cannot be answered without assessing the outcomes of the prescribed fire and comparing pre- and post-fire conditions. If the prescribed fire was not effective in realizing the management goals, then the fire plan should be amended accordingly, for example, by decreasing or increasing the frequency or intensity of burns. Effective monitoring provides the basis for making an objective evaluation and tracking courses of action and, when nec-

essary, modifying the management plan. The response of a target wildlife species to a fire might entail many years of monitoring, given the
lag time between the fire, benefits of the fire (e.g., food production),
and the response of the wildlife species (e.g., density, productivity).

Mechanical Treatments

Mechanical treatments (Aldo Leopold's "axe") comprise a broad suite
of methods used to control vegetation by physically damaging, uprooting, or removing plants. Some examples of mechanical treatments are
a tractor pulling a disk or roller chopper or the use of hand tools like a
chainsaw or axe. Regardless of the level of complexity of the tool being
used, the effects of mechanical treatments are simply to manipulate
vegetation and/or modify soil conditions for a given management goal.
We describe mechanical treatments in two broad categories: (1) treatments where the target vegetation is *merchantable* (has some mone-
tary value) and harvested for the production of wood products, and (2)
treatments where the target vegetation is not typically harvested and is
instead destroyed or manipulated for the benefits of early-successional
plant communities (e.g., midstory, understory removal). Mechanical
treatments in the latter category often are described as "brush management," a term we use to describe mechanical treatments typically
applied to vegetation less than 6 meters (20 ft) in height comprising the
midstory/understory layers within a plant community. Many times one
or more of these mechanical treatments are used from both categories
depending on the management objective and purpose of the treatment.

Mechanical Treatments: Midstory/Understory

This section provides tools to assist managers in determining when to
use mechanical treatments, assess which mechanical equipment is the
most effective and appropriate for the specified goals and objectives,
and develop and implement an effective management plan when dealing with mid- and understory vegetation. Mechanical treatment is
another land management tool used to control vegetation by physically
damaging, uprooting, or removing plants. Mechanical tools such as
tractors, roller choppers, and chainsaws are used to manipulate vegetation and/or modify soil conditions. In addition, mechanical treatments
are used to reduce hazardous fuels, reduce water runoff, and break up
compacted soils (Haufler and Ganguli 2007). Hence, mechanical treatment methods are integrated land management practices used to protect plant communities, natural resources, ecosystems, and habitats.
Mechanical tools are also used to remove nuisance vegetation or inva-

sive plants. In addition, mechanical treatments are used to improve soil permeability and increase water infiltration to increase production of herbaceous grasses, forbs, sedges, and legumes.

Function and Benefits

The beneficial effects of mechanical treatments are generally short-lived but in many cases are cost-effective, as the range of applications includes individual plant removal (with axes, machetes, etc.) or the use of heavy machinery for quickly clearing large tracts of land (Scifres and Hamilton 1993). Nonetheless, mechanical treatments alter plant community composition and modify rangeland production biomass (Miyamoto, Olson, and Schuman 2004).

Different mechanical treatment methods have varying effects on plant communities. Climatic conditions, soil properties, and other environmental factors significantly impact how brush and other undesirable vegetation are managed. Understanding vegetation root systems, soil properties, and climatic conditions will determine the best mechanical treatment option to effectively and properly manage varying plant communities. Table 4.3 provides a list of commonly used mechanical treatments and fundamental information on application to assist in determining the most effective mechanical method.

Application

Mechanical treatments can be hand- or power-applied methods. Both methods vary in cost and application technique and time. The beneficial effects are determined by the plant community, species density, selected method, terrain, and climatic conditions. Hand treatments such as cutting, girdling, and grubbing are very effective, but their application is often time consuming. Power methods (chaining, roller chopping, etc.) are less labor intensive but often have high equipment costs. However, different mechanical treatments are used to target different plant communities, and vegetation response varies according to soil properties, the existing plant community, climatic conditions, rainfall, and prescribed method. To achieve specific goals and results, land managers often administer mechanical treatments several times to control invasion/reinvasion of undesired plants. Also, mechanical treatments are often used in conjunction with other rangeland management tools such as prescribed fire and application of chemicals. When properly administered, these treatments cause minimal damage to desirable plants and animals while controlling undesirables. It is important to note that the combination of two or more brush manage-

Table 4.3. Commonly used mechanical treatments and applications

Mechanical method	Equipment	Application	Limitations	Percent of plant mortality rate
Dozing	Bulldozer	• Land clearing by pushing, pulling, and removing brush, trees, and stumps <6 inches in diameter	• Trees and stumps >6 inches require special methods and take longer to remove by dozer • Maneuverability limited, especially in wet or muddy terrain and dense growth	>95% of aboveground ground <6 inches in diameter
	Tree-dozer/ tractor-mounted unit/plow	• Clearing medium-size brush and trees at ground level	• Skilled equipment operator for rigging • Maneuverability limited in swampy, muddy terrain • Slow and time-consuming process	>95% of aboveground ground stems <10 inches in diameter
	Tractor-mounted clearing units	• Extensive clearing operations such as uprooting any size tree or stump • Can be used in swamps, bottomlands, and jungles with any size vegetation growth	• Skilled equipment operator for rigging • Slow and time-consuming process	>95% of vegetation cover
Root plowing	Tractor-mounted plows	• Extensive clearing operations such as uprooting any size tree or stump • Loosens soil for stripping or moving surface boulders	• Skilled equipment operator for rigging • Slow and time-consuming process	>95% of vegetation cover
	Winches (tractor or truck mounted)	• Tractor mounts pull or uproot trees or stumps up to 24 inches in diameter • Truck mounts pull small trees up to 6 inches in diameter	• Pull capacity limited by tractor/truck horsepower • Maneuverability limited by terrain • Rigging personnel required	>95% of aboveground stems <6 inches in diameter
Sawing	Chainsaw	• Rapid felling of trees • Trunk size unlimited	• Uprooting stumps requires different equipment • Pneumatic saws dangerous in steep and rugged terrain • Requires skilled operators	>95% of vegetation cover
	Circular or chainsaw mounted on tractor	• Rapid felling of any size tree • Excellent for clearing heavy, dense growth in rough terrain	• Uprooting stumps requires different equipment • Maneuverability limited in swampy, muddy terrain • Blades may bind	>95% of aboveground stems

Mechanical method	Equipment	Application	Limitations	Percent of plant mortality rate
Ripping	Ripper	• Cuts tree roots • Loosens soil for stripping or moving surface boulders	• Maneuverability limited in swampy, muddy terrain or dense growth • Depth of shank limits use to shallow roots	
Shearing	Shears	• Uprooting stumps requires different equipment • Pneumatic saws dangerous in steep and rugged terrains • Requires skilled operators	• Skilled operator required	>95% of aboveground stems <6 inches in diameter
Grading	Grader	• Clears grass, weeds, or small brush/vegetation. • Grades drainage areas	• Maneuverability limited to level, even terrain free of trees, stumps, brush, or boulders • Skilled operator required to prevent damaging blades	
Pulling/ grubbing	Grubber excavator	• Extensive clearing operations such as uprooting any size tree or stump	• Maneuverability limited in swampy, muddy terrain or dense growth • Skilled operator required	>95% of aboveground stems
Brush mower	Mower	• Clears grass, weeds, or small brush/vegetation • Grades drainage areas	• Maneuverability limited to level, even terrain free of trees, stumps, brush, or boulders • Skilled operator required	>95% of aboveground stems <3 inches in diameter
Girdling	Girdling knives, machetes	• Extensive clearing of any size tree or stump	• Skilled knowledge and application technique necessary to avoid damaging the xylem vessels	>95% of top growth above girdle

ment methods is always more effective and longer lasting than a single method and sometimes more economical as well (more time between treatments due to the longer-lasting results; Bovey 1998).

The application of mechanical treatment requires specific objectives in accordance with land uses and livestock and/or wildlife grazing. Selecting an appropriate method should be done after analyzing projected outcomes measured against possible disadvantages (e.g., soil erosion, reduced runoff to reservoirs, restricted vehicle use, invasion of unwanted vegetation). Other factors for consideration include travel lanes, loafing areas, and escape cover for livestock and wildlife. Furthermore, wildlife species vary in their vegetative structural needs and, like cattle, dislike crossing over furrows (Natural Resources Conservation Service 2001). Landscape variances such as terrain and topography should also be considered, especially if mechanical treatment options are limited. For these reasons, it is important to be objective when selecting mechanical treatments, particularly because of the varying costs and application methods.

Dozing is very effective in controlling brush or larger trees by pushing or removing plants by the roots (fig. 4.4). Dozers can have a straight blade, blades with teeth, or V-shaped "stinger" attachments. This tool is often used on rocky soils or open areas with a stand of trees.

Chaining is the process of pulling each edge of a chain (with 36+ kg, or 80+ lb links) by two tractors on opposite ends of the chain (fig. 4.5). This method is very effective on single-stemmed trees that are at least 0.3 meter (1 ft) in height. Chaining is an economical mechanical process.

Girdling is the removal of bands of bark and the underlying cambium and phloem in a band around a tree (fig. 4.6). This process cuts off or reduces the flow of nutrients, causing the tree to slowly die. Also, it is important not to damage the xylem vessels during girdling, which will cause the xylem to send signals to the tree to send up a shoot. Girdling is

Figure 4.4. Bulldozing. Image by Joyce VanDeWater.

Figure 4.5. Chaining. Image by Joyce VanDeWater.

Figure 4.6. Girdling. Image by Joyce VanDeWater.

not effective on resprouting species because growth is from the stump or root system. The most effective method requires the inner bark to be continuously removed from around the tree in a band 2.5 centimeters (1.0 in) or wider.

Grubbing is the process of pulling or cutting unwanted vegetation by hand or with mechanical tools. Grubbing machinery such as an excavator can uproot vegetation from the soil and remove or translocate it.

Railing and *cabling* are similar to chaining and are traditionally used to remove small, limber trees and shrubs located in dense stands. However, railing, cabling, and chaining are sometimes ineffective brush-control processes due to riding-up (going over target vegetation), skipping, or missing smaller plants.

Shearing and *roller chopping* chop or reduce/diminish plants in size. Shearing removes the aboveground portion of plants and is very effective in removing overstory and midstory growth. Roto-beaters and roller choppers are better at removing understory and woody plants that are less than 10 centimeters (4 in) in diameter. However, use of these tools on resprouting vegetation will only temporally suppress plant growth.

Disking is usually limited to small, shallow-rooted shrubs and is conducted with a large plow or tandem disk to uproot brush. Disked areas should be seeded to establish desirable plant species, minimize erosion, and prevent recolonization by undesirable species.

Mowing and *shredding* are methods to control small-stemmed brush and weeds. These types of treatments require repeated application for effective control.

Root plowing is the process of pulling a horizontal blade with a tractor to cut brush 46–61 centimeters (18–24 in) belowground (fig. 4.7). This process has the highest plant mortality rate of any other method and is very effective in areas of mixed-brush species. Areas should be seeded after root plowing to establish desired species, prevent undesired species, and minimize erosion.

Axes, *girdling knives*, *hatchets*, *saws*, *shears*, and *shovels* are used to manually remove or uproot vegetation by cutting limbs, trunks, or bands of bark or digging out plants by their roots. These mechanical tools are effective but time consuming. However, these tools are often employed in areas where motorized or heavy equipment cannot be used.

Figure 4.7. Root plowing. Image by Joyce VanDeWater.

Planning

The state of the mid- and understory components of forests impacts plant structure, fire frequency and intensity, hydrology, soil chemistry, economics (e.g., timber, recreation), and biodiversity. Mechanical treatment of the midstory/understory is a direct management strategy that demands careful planning and has great impact. Poor planning has resulted in degraded rangelands and forests full of undesirable plant and wildlife assemblages that negatively impact ecosystem function (Fulbright 1996). Failures in ecosystem management, including poor mechanical treatment strategies, often have impacts obvious to the public (e.g., vegetation changes, water quality, wildfires). Managers must now expect an increasingly critical awareness by the public. Chainsawing, dozing, and other such techniques must be well planned, justified by need, but also explained to stakeholders. The differences between destroying a habitat with a dozer and appropriately managing that habitat with a dozer are not always readily apparent to the observant public.

Managers must begin with clear goals supported by understanding of the ecosystem and experience with mechanical treatments. This is often dependent on the scale of the operation (e.g., single small riparian area or an entire watershed), other management actions (e.g., wildlife management, resource extraction), species treated, other impacted species, location, and operational budget. Such actions can impact a variety of species to large degrees (Provencher et al. 2002). Mechanical treatments can result in similar outcomes to those of other prescribed disturbance regimes but rely on knowledge of appropriate species removal, vegetation density, and impacts of management actions such as soil compaction and erosion and employ comprehensive habitat- and species-monitoring techniques. As necessary for other management options, mechanical treatments require knowledgeable and experienced managers with the commitment to exhaustively monitor treatment areas prior to and after management actions.

Analyzing and evaluating landscapes, understanding vegetation dynamics, and identifying specific land management goals and objectives are critical and essential steps in developing a mechanical treatment management plan. Recognizing and outlining safety measures as well as identifying potential risk factors associated with various types of equipment are also fundamentally important steps in planning and implementing such plans. Other planning considerations include timing of application, climatic conditions, topography of target area, and specialized mechanical operational skills.

A thorough understanding of target plant biological growth and regrowth functions is essential when planning mechanical treatment methods to ensure that the correct machine and method are used and properly applied. Well-thought-out vegetative management plans always incorporate short- and long-term goals before mechanically treating a target area. Mechanical treatment plans also identify wildlife habitat needs and often combine a variety of treatment methods for maximum effectiveness and reduce habitat management operational costs. Novice land managers should consider seeking assistance from county extension agents/specialists or NRCS technicians when planning mechanical treatment management plans and should ask for mechanical treatment application recommendations.

The mechanical treatment plan checklist provided in table 4.4 serves as a planning guide. It does not contain a comprehensive list of mechanical treatment options but provides important areas to be considered.

Costs

The cost of applying mechanical treatments will vary according to total area of treatment, terrain, complexity of treatment pattern, current fuel prices, and contracting machinery operators and service fees. These costs are less expensive when operating equipment is owned or inexpensively rented. In general, mechanical treatment cost factors include required equipment, modifications necessary to enable equipment to adequately complete the job, follow-up treatments, fuel, season or weather conditions on projected treatment dates, liability insurance and permit fees if applicable, and contracting services. Costs can be reduced by using local contractors, who have less travel/transportation expenses. Such contractors are likely to be listed under "commercial vegetation treatments." Table 4.5 provides a general list of potential cost factors.

Implementation

It is important for managers to first determine if the application of mechanical treatment is needed to meet land management goals and objectives. Second, managers must consider other land management options or a combination of habitat management methods, for instance, prescribed fire and mechanical treatment to achieve the desired land management goals for wildlife and/or livestock. If mechanical treatment is determined to be the optimal land management solution, the risks and costs should be weighed against the short- and long-term benefits. Land management plans also need to be reviewed and revised as necessary to obtain desirable projected outcomes.

All mechanical equipment should be inspected and be in optimal working condition before application. Equipment should also be used for the recommended mechanical treatment and/or retrofitted for the specific task; failing to do so may cause injury to the operator, damage to habitats and ecosystems, or damage to equipment. Weather conditions should also be optimal to reduce risk and injury. The plan specifics should be outlined and discussed ahead of time and/or the day of the treatment. The site location, terrain, acreage, and method will determine the course of action. Access to treatment area(s), vegetative type, and time of the year will also determine a strategic course of action and whether the treated area needs to be reseeded.

Monitoring and Assessment

Monitoring, evaluating, and assessing application of mechanical treatments reduce the wildlife habitat risk factors and ensure that desired measurable outcomes are accomplished. Whenever mechanical treatment management goals are not achievable in accordance with the objectives, the application plan should be reviewed and revised accordingly. Effective monitoring and time-appropriate assessment of mechanical treatment facilitates tracking the treatments allows time for necessary modifications to the plan. The impact to wildlife habitat and food and water sources should also be monitored and assessed on a regular basis to minimize negative effects.

Mechanical Treatments: Overstory

In this section we review mechanical treatments commonly used in management of forests (also known as *commercial woodlands* in the eastern and southeastern United States), the overstory, or canopy layer. Future managers should understand that management is often conducted within forestry activities. Wildlife professionals in the southeastern United States or in the Pacific Northwest, for example, typically work with forest professionals in the integration of wildlife habitat recommendations. We begin with some basic definitions. *Forestry* is the science and art of creating, managing, and using forests and associated resources in a sustainable manner. *Silviculture*, a related science within forest management, is the practice of controlling the establishment, growth, and composition of forests to meet various human needs and values. The application of silvicultural practices often is applied to a *forest stand*, a group of trees of sufficiently uniform species, composition, age, and condition to be considered a homogeneous unit for management purposes. The application of silvicultural practices can be broken into (1) regeneration treatments and (2) intermediate stand treat-

Table 4.4. Mechanical treatment checklist

Pre-treatment rationale
> Identify and target pastures
> Purpose (reclamation, maintenance, etc.)
> Timing (cool or warm season)
> Consideration of vulnerable areas (erosion, wildlife, etc.)
> Legal responsibilities (liability)

Planning mechanical treatments (planning is conducted 6 months to a year in advance)
> Consider alternative feeding locations for livestock, wildlife, etc.
> Identify potential treatment dates at least 12 months in advance
> Identify budget cost
> Identify treatment application dates
> Identify treatment methods
> Identify and list mechanical treatment tools and whether trained machinery operators will be needed or if services will be contracted out
> Complete a comprehensive mechanical treatment management plan

Post-treatment considerations
> Reseed area if applicable and monitor vegetation changes
> Check for livestock and wildlife access
> Observe vegetation changes
> Identify and evaluate when treatment areas will reach optimal conditions to restock livestock (calculate proper stocking rates)
> Project grazing periods and monitor changes

Mechanical treatment action plan

Identification of treatment site (must be completed 6 to 12 months in advance)
> Area location
> Area size (in acres)
> Proposed date(s) (12 months in advance)
> Set date(s) (6–8 months in advance)
> Outline perimeters of treated areas; also sketch area with as much detail as possible, i.e., include topographical information when possible

Identify and list land management and treatment objectives

Fuel reduction	Woody plant control
Weed control	Forage improvement
Increased wildlife habitat	Disease control
Improved wildlife diversity	Reduce understory/overstory growth
Other	

Site characteristics
> Percent of grasses, forbs, and legume cover
> Percent of low-volatile plant cover (e.g., honey locust)
> Percent of high-volatile plant cover (e.g., eastern redcedar)
> Percent of leaf matter and other debris
> Identify slope and terrain of the site

Level terrain	Level to rolling terrain
Moderately steep	Steep

> Identify and list all potential hazards near the perimeter of the treatment site (electrical lines and poles, gas wells, barns, etc.)

Identify and list bodies of water, points of access, location of firebreaks, etc.

Area location

Area size (in acres)

Proposed date(s) (12 months in advance)

Set date(s) (12 months in advance)

Outline perimeter of treated areas; also sketch area with as much detail as possible, i.e., include topographical information when possible, indicate firebreaks and escape routes

Identify and list land management and treatment objectives

Fuel reduction	Woody plant control
Weed control	Forage improvement
Increased wildlife habitat	Disease control
Improved wildlife diversity	Reduce understory/overstory growth
Other	

Acceptable weather and moisture conditions

Wind velocity	Wind direction
Relative humidity	Temperature
Soil and plant cover moisture	Other

Apply for treatment permits (if applicable)

Obtain contact information, i.e., name of agencies, address, phone number, website, contact personnel's e-mail address, and date permit was issued

Notify state and local agencies	Address, telephones, websites, e-mail addresses	Contact personnel

Equipment requirements

	Item description	Model number	Source
Mechanical equipment			
Safety equipment			
Other			
Equipment operator contact information	Address, telephone numbers, websites, e-mail addresses	Contact personnel	
Emergency and first-responder assistance	Address, telephone numbers, websites, e-mail addresses	Contact personnel	
Firefighter			
Red Cross			
Emergency first responders			
Other			

Mechanical treatment methods

Plow	Disking	Grubbing
Mowing/shredding	Chaining	Root plowing
Dozing	Railing	Cabling
Roller chopping	Shearing	Other

Date of mechanical treatment

Source: Adapted from Seifres and Hamilton (1993); Volesky, Stubbendieck, and Mitchell (2007).

Table 4.5. Mechanical treatment cost checklist

Fuel	**Grazing loss cost (lease rate)**
Diesel	**Labor cost**
Gasoline	Number of personnel x daily rate
Application cost	**Other labor-related expenses**
Mowing/shredding	Travel expenses
Disking	Per diem
Dozing	Overnight accommodations
Plowing	**Protective gear**
Grubbing	Hats, gloves, etc.
Chaining	Eye and ear protection
Root plowing	**Drinking water**
Railing/cabling	**First-aid kit**
Shearing	**Liability insurance (if applicable)**
Roller chopping	**Contracted services**
Other method	**Other miscellaneous expenses**
	Total cost
	Total acres treated
	Total cost/acre

ments. The focus of our discussion is primarily on the basic concepts of silviculture and less on the mechanical treatment itself because many of the tools used in forest management are similar to those described in the previous section.

Function and Benefits

There are many benefits to mechanical treatments used as an integral part of forest management, including (1) establishment of new forest stands, (2) increase of tree growth or biomass through the reduction of competition, (3) modification of forest composition through targeted species selection, and (4) improvement of overall forest health. Sustainable forestry often requires the active management of forest stands to ensure the successful reestablishment of plant communities following a harvesting operation. Forest activities that do not include reestablishment of preferred plant communities are considered logging rather than sustainable forestry. Mechanical treatments are critical to ensuring that these new forest stands are established through several approaches collectively referred to as *regeneration systems* (discussed in the implementation section). The benefit of these treatments is obviously the regrowth of preferred plant communities. For example, early-successional forest species like southern pines (*Pinus* spp.) require mechanical treatments that favor the species in reestablishment.

A second benefit of mechanical treatments is the reduction of both inter- and intra-plant species competition. Reduction of plant density serves to "release" favored tree species, resulting in increased growth or biomass/volume. This leads to an increase of merchantable tree volume (wood products). Mechanical treatments also modify forest composition through selectively targeting or favoring tree species and removing unwanted species. Finally, use of mechanical treatments maintains lower tree densities within a forest stand, which typically makes trees less susceptible to forest pests such as the southern pine beetle, because many forest pests and pathogens are density-dependent.

Application

Application of mechanical treatments is complex and dependent on set goals and manager experience. Managers ultimately want functioning ecosystems. For instance, midstory and understory mechanical treatments in forests are not designed to eliminate the midstory/understory. Generally, managers want a mixture of overstory, midstory, and understory in a sustainable state that benefits management goals (e.g., increased biodiversity, increased timber extraction). These mixtures can often be attained through a combination of different management options, including mechanical, herbicidal, and prescribed fire treatments. For example, spot herbicidal treatments might immediately follow mechanical removal of midstory trees to minimize regrowth. Mechanical treatment is also employed prior to some prescribed burns to increase burn efficacy or control. Managers must also consider season of treatment, location, and personnel experience/training. Applying mechanical treatment during multiple seasons can spread impact to different species at different times. This may improve management impact and success. Personnel experienced with heavy equipment and vegetation control can improve treatment impact, lower costs, and reduce time investments.

Planning

Proper forest planning is an important aspect of forest management. Many of the approaches used in modern-day forest management that include mechanical treatments are conducted within a general set of guidelines known as BMPs, introduced earlier in this chapter. These forestry guidelines are important to ensure the proper implementation of management activities to minimize potential negative environmental impacts. Most state forestry agencies, or federal entities if work is being conducted on federal lands, have forest BMPs available for land managers to consider during their planning processes. We refer the reader to these important references to review and consider during the forest planning process.

The timing of mechanical treatments in sustainable forest management is influenced for the most part by the accessibility to a forest stand. Ground conditions are a major factor. Forest harvesting or the implementation of intermediate forest practices, such as tree thinning, are generally not recommended when ground conditions are too wet or when heavy machinery typically used in conducting these operations would cause environmental degradation to a site. As previously mentioned, BMPs typically outline these various practices to minimize potential negative impacts. Avoidance of sensitive time periods such as nesting seasons should be considered when using these mechanical treatments.

Forests have long been recognized as providing public benefits and services such as clean water, clean air, and wildlife habitat beyond just wood production. Forest BMPs were developed as standards in all states with an active forest industry as a result of historic issues related to forest practices and their impact on water quality and quantity. The Clean Water Act of 1972 further solidified the requirement for standards to prevent sediment from entering waterways during harvest operations and other forest management activities. Types of forest BMPs include implementation of stream buffers, avoidance of wet-weather logging, and proper road design. Forest BMPs related to mechanical treatments might include guidelines for constructing access roads and skid trails (e.g., avoid natural drainage areas, outslope roads, and ditches to remove water from roadway), leaving undisturbed filter or buffer strips of trees and vegetation (streamside management zones [SMZs]), and stabilizing the soil on log landings and skid trails following harvest by seeding with grass or legumes.

Costs

The majority of costs for mechanical treatments used during forestry operations are often offset by income generation from timber harvests and other commercial activities such as forest thinnings. Operational costs are relatively expensive due to the specialized heavy machinery required. These costs are factored into the long-term economics of forest plans. In essence, forestry involves a long-standing "crop" in the form of trees that typically are harvested on long rotations (20+ years). For these reasons, the costs associated with mechanical treatments are factored into the long-term economics of a given forest stand.

Implementation

Implementation of mechanical treatments in forest management can be broken into (1) regeneration treatments and (2) intermediate stand treatments. *Forest regeneration* is the process of renewing tree cover by establishing young trees naturally or artificially after the previous stand or forest has been removed. The methods of regenerating a forest stand can be categorized into six classes. Single-tree selection and group selection result in a forest stand that is uneven in age or has a tree composition of various age classes. Clear-cutting, seed trees, shelterwood, and coppicing are approaches that result in an even-aged forest stand.

Single-tree selection is an uneven-aged regeneration method most suitable when shade-tolerant species regeneration is desired. It is typical for older and diseased trees to be removed, thus thinning the stand and allowing for younger, healthy trees to grow. Single-tree selection can be very difficult to implement in dense or sensitive stands, and residual stand damage can occur.

Group selection is an uneven-aged regeneration method that can be used when shade- to partial-shade-tolerant species regeneration is desired. This method can still result in residual stand damage in dense stands; however, directional felling can minimize the damage. Additionally, foresters can select across the range of diameter classes in the stand and maintain a mosaic of age and diameter classes.

Clear-cutting is an even-aged regeneration method that can employ either natural or artificial regeneration. Clear-cutting can be biologically appropriate for species that typically regenerate from stand-replacing fires or other major disturbances, such as lodgepole pine (*Pinus contorta*). Alternatively, clear-cutting can change the dominating species on a stand with the unintended introduction of nonnative and invasive species. It can also prolong slash decomposition, expose soil to erosion, impact visual appeal of a landscape, and remove essential wildlife habitat. It is particularly useful in regeneration of tree species such as Douglas-fir (*Pseudotsuga menziesii*), which is shade intolerant. State-level forest practice rules usually require that, following clear-cutting, trees be reestablished within a certain time frame. Seedlings are often hand-planted, which also allows foresters to change the native tree-species composition to one made up predominantly of more commercially viable species.

Seed tree is an even-aged regeneration method that retains widely spaced residual trees to provide uniform seed dispersal across a harvested area. In the seed-tree method, 5–30 seed trees per hectare are left standing to regenerate the forest. They are retained until regeneration has become established, at which point they may be removed. It may not always be economically viable or biologically desirable to re-enter the stand to remove the remaining seed trees. Seed-tree cuts can also be viewed as a clear-cut with natural regeneration and thus can have all of the problems associated with clear-cutting. This method is most suited for light-seeded species and those not prone to windthrow (trees uprooted or knocked down by wind).

Shelterwood is a regeneration method that removes trees in a series of three harvests: (1) preparatory cut, (2) establishment cut, and (3) removal cut. The objective is to establish new forest reproduction under the shelter of the retained trees. Unlike what occurs in the seed-tree method, residual trees alter understory environmental conditions (sunlight, temperature, and moisture) that influence tree seedling growth.

Coppicing is a regeneration method that depends on cut trees' sprouting. Most hardwoods, the coast redwood, and certain pines naturally sprout from stumps and can be managed through coppicing, which is generally used to produce fuelwood, pulpwood, and other products dependent on small trees. In compound coppicing or coppicing with standards, some trees of the highest quality are retained for multiple rotations to obtain larger trees for different purposes. A close relative of coppicing is pollarding (pruning top of trees to promote denser sprouting at a lower height).

Intermediate stand treatments are forest operations used to enhance or expedite the growing forest stand to the end of its cycle or rotation (the time when a forest is harvested and the process repeats). Examples of intermediate stand treatments include stand thinning (reduction of tree density to stimulate the growth of more dominant trees in the forest canopy) or even precommercial thinning, which involves the removal of subdominant trees but without generation of revenue. Generally precommercial thinnings are done when trees are too small for harvest of any wood product. Silvicultural regeneration methods combine both the harvest of the timber on the stand and reestablishment of the forest.

The proper practice of sustainable forestry should mitigate the potential negative impacts, but all harvest methods will have some impact on the land and residual stand. The practice of sustainable forestry limits the impacts such that the values of the forest are maintained in perpetuity. The methods by which managers spur forest growth (release) are (1) weeding, a treatment implemented during a stand's seedling stage that removes or reduces herbaceous or woody shrub competition; (2) cleaning, release of select saplings from competition by overtopping trees of a comparable age, which favors trees of a desired species and stem quality; (3) liberation cutting, a treatment that releases tree seedlings or saplings by removing older overtopping trees.

Thinning is the reduction of tree density to stimulate the growth of more dominant trees in the forest canopy. The goal is to control the amount and distribution of available growing space. By altering stand density, foresters can influence the growth, quality, and health of residual trees. Thinning also allows for the removal of less commercially desirable trees such as small, deformed, or diseased individuals. Unlike regeneration treatments, thinnings are not intended to establish a new tree crop or create permanent canopy openings. These are common thinning methods:

- Low thinning (thinning from below, or German thinning)
- Crown thinning (thinning from above, or French method)
- Selection thinning (thinning of dominants, or Borggreve method)
- Mechanical thinning (row thinning or geometric thinning)
- Free thinning (combination of low and crown thinning)

Ecological thinning is carried out when the primary aim of forest thinning is to increase growth of selected trees, favoring development of wildlife habitat such as hollows rather than focusing on increased timber yields. Ecological thinning is also considered a new approach to landscape restoration for some types of eucalypt forests and woodlands in Australia.

Monitoring and Assessment

Mechanical treatments must be carefully reviewed to ensure that desired ecosystem changes take place. Monitoring mechanical treatment as it occurs and assessing immediate and long-term success provide valuable insight for managers. As mechanical treatments in the midstory/understory are generally designed to reduce a species or suite of species for the benefit of others, only careful pre- and posttreatment monitoring can assess success. Additionally, posttreatment monitoring can alert managers to areas that require follow-up treatments or management actions that must be altered. Careful consideration of treatment location and extent followed by assessment can also provide a guide for future management actions. We must also remember that mechanical treatments are often just one step in a multistep management strategy (often combined with herbicides, prescribed fire, etc.). Thus, mechanical treatments must meet certain minimum goals for other treatments to succeed. For example, insufficient cutting of mid-story trees might make planned herbicide treatments ineffective or impossible.

Monitoring and assessment can require more than simply determining whether mechanical treatment removed the target vegetation. It often requires managers to monitor other impacts such as preferred vegetation growth and wildlife species population changes. Consider wildlife management goals such as encouraging the recovery efforts for the endangered red-cockaded woodpecker (*Leuconotopicus borealis*), a species closely tied to specific forest structure. Extensive monitoring and assessment are required before, during, and after treatment. It is critical for managers to remember that management actions must have goals and be evaluated in that context.

Grazing

Grazing (Aldo Leopold's "cow") is yet another management tool. We provide information in this section to elevate a manager's understanding of scientific BMPs related to grazing that can improve, maintain, and sustain habitats. We discuss the mechanics of grazing domestic livestock, evaluate and assess the various grazing systems, and determine which grazing systems will yield the desired land management goals and objectives.

Grazing lands, over time, have evolved as a result of the various interactions between herbivory, carcass decomposition, heavy soil disturbance, and nitrogen deposits. These grazing forces have changed both ecosystem function and vegetation communities. In general, grazing behavior affects species composition when herbivores select or avoid specific plants. This grazing pattern, known as *selective grazing*, reduces competitive vigor of grazed plants and increases ungrazed plant species competition. Unfortunately, poorly managed lands often have increased invasive plant populations as well as reduced vegetative cover and compact eroded soils.

The concept behind grazing as a management tool is to shift grazing intensity of specific areas such as pastures by rotating livestock to better utilize plant communities and increase animal production/animal cover. Grazing is also used to clear land, control invasive or unwanted vegetation establishment and growth, or improve specific animals' habitat. Grazing management goals and objectives require extensive knowledge of plant response to grazing behavior. Periodic rest periods for specific areas/pastures are also essential management practices of grazing. In addition, grazing of vegetative-sensitive areas should be avoided and properly managed for vegetative restoration. It is also

Pasture Productivity

Description: Rotational grazing moves grazing animals from one pasture to another over time to allow some pastures to rest from grazing pressure. Rotational grazing can increase per-animal profits and reduce negative range impacts if managed correctly. Proper rotational grazing is a relatively complex endeavor and requires that the manager understand range ecology, livestock needs, and pasture productivity. Estimating pasture productivity can be somewhat daunting but is not terribly complex. There are several methods for estimating pasture productivity, but we focus on a single method: direct estimate method (Undersander et al. 1992). In this method, the manager uses scissors or shears to collect forage in 1 square yard of pasture at the grazing height. This sample is weighed in pounds. A small subsample is collected from this sample, weighed, then dried using a vegetation dryer or microwave. The dry matter is then weighed and used in the final calculations.

$$\text{Percent forage dry matter} = \frac{\textit{Dry subsample weight}}{\text{initial subsample weight}}$$

$$\text{Pasture yield (lb/ac)} = \frac{[(\textit{Initial forage weight}) \times (\textit{Percent forage dry matter}) \times (43{,}560 \; \textit{ft}^2\textit{/ac})]}{(9\text{ft}^2\text{/yd}^2)}$$

Scenario: The direct estimate method as detailed by Undersander et al. provides a pasture forage yield in pounds per acre. Imagine that you have already collected and prepared the samples from a pasture you are managing. Determine the productivity of your pasture using the above calculations. What is this final productivity number telling you?

Initial forage weight = 1.76 lb/yd^2
Initial subsample weight = 7 oz
Dry subsample weight = 1.3 oz

important to understand livestock's dietary preference and grazing selection. In general, grazing management applications focus not only on the kind and class of grazing livestock but also on stocking rates, land carrying capacity, and plant responses.

Function and Benefits

Grazing management methods require managers to evaluate the interactions between wildlife and livestock grazing behavior, vegetation preference, and land carrying capacity. Understanding grazing dynamics minimizes overgrazing and allows livestock stocking rates to be adjusted, thereby decreasing wildlife herbivory competition. Grazing functions are also important in regard to the protection of fragile ecosystems and habitats, as these areas need to be monitored and protected to restore vegetation.

Assessing the grazing merit or value of the vegetation community determines palatability and animal preference. Therefore, land managers who raise livestock can establish and manage healthy ecosystems that can support livestock and wildlife. Sustainability of natural resources depends on properly managing water, soil, vegetative communities, and grazing activities.

Application

Grazing systems are traditionally characterized as natural, domesticated, or prescribed. Every grazing system entails an array of grazing options that ultimately affect not only the plant composition but also the nutritional status of animals. In grasslands, as the grazing season progresses, the protein content of grasses, forbs, and legumes decreases and drops below the requirements of animals (Cook and Harris 1968). However, in forested areas, the nutrient value increases as the grazing season progresses (Walburger et al. 2000). Identifying measurable grazing goals and understanding grazing techniques are key factors in successful grazing applications and should be analyzed within the context of stocking rates, forage quality, terrain, climatic conditions, duration of grazing season, and availability of water.

Natural grazing systems allow native ungulates (hoofed animals) to graze on a variety of plants traditionally for short periods of time and over large geographic areas. For example, migratory bison (*Bison bison*), elk (*Cervus canadensis*), pronghorn (*Antilocapra americana*), white-tailed deer (*Odocoileus virginianus*), and mule deer (*O. hemionus*) harvest native grazing vegetation. The population density of ungulates is greatly determined by natural population-control factors (disease, forage production, predatory animals). Ungulates have coexisted over 10 million years and play an important role in maintaining plant species composition and ecosystem function (Frank, McNaughton, and Tracy 1998). As a result, plant communities have adapted to frequent and short-duration grazing.

Domesticated grazing systems utilize pasture and rangelands for production of domesticated livestock (Frank, McNaughton, and Tracy 1998; Freilich et al. 2003). Unlike sustainable vegetation of natural grazing systems, domesticated grazing systems require vegetation management strategies to maintain viable forage production to support livestock production. There are four basic grazing methods: (1) deferred rotation, (2) rest rotation, (3) high intensity–low frequency; and (4) short duration (which requires two pasture systems: rotational stocking and continuous stocking). However, other systems incorporate variations and combinations of these pasture systems.

Rotational stocking systems (rotational grazing) require pasturelands to be subdivided into paddocks into which livestock are rotated to graze intensively for short periods of time. These systems can be based on simple or intense rotation, and deferred grazing periods should be adequate (can last up to two years) to allow restoration of the plant community. The ultimate goal is to force livestock to maximize grazing on forage before relocating the stock to another paddock. Hence, these systems allow livestock to be moved based on forage growth to promote better forage utilization; thus, grazing seasons are extended, and grazing of specific pastures is deferred.

Paddocks can be fenced with portable fencing materials and should be sized according to the number of livestock, vegetation composition, daily forage intake (grazing pressure), and number of grazing days. Other factors that affect paddock size are land productivity functions, climatic conditions, and availability of water. Fencing can also be used to protect vegetatively sensitive and riparian areas, wetlands, and other ecologically sensitive areas. If water is not naturally accessible, supplemental water can be provided via water tanks, pipelines, windmills, and so forth. Unfortunately, human-managed grazing systems cannot mimic natural grazing patterns by native herbivores (Freilich et al. 2003).

Continuous stocking systems (continuous grazing) allows livestock to remain in the same pasture area for an undetermined period of time, and grazing is not restricted anytime during the grazing season. If numbers of livestock remain the same, then this method is called *set stocking* (Clark Conservation District 2013). Continuous stocking systems are the most common type in the United States because of low investment cost. However, over time these systems increase less desirable plant species since lack of grazing restrictions allows livestock to consume the most palatable forage repeatedly so that it recovers slowly or not at all. Eventually, the less desirable/palatable or nonpreferred plant species mature and increase their establishment, while preferred species are reduced or eliminated. In addition, the potential for invasion of undesirable species increases along with soil degradation and soil erosion. The ultimate outcome is poor forage quality and reduced animal performance. Trampling and livestock fecal matter further reduce forage utilization and animal production. Continuous grazing is most successful when grazing livestock are of moderate to low milking production (milking animals demand greater forage intake) and stocking rates are maintained at the adequate carrying capacity rate throughout the year.

Prescribed grazing integrates the use of livestock to reduce undesirable vegetation to give native vegetation a competitive edge. Prescribed grazing systems are designed to cause maximum damage to target vegetation while reducing excessive damage to desirable plant communities. Prescribed grazing is also used to control fuel-loading vegetation in high-risk locations, such as areas experiencing extreme drought. Successful prescribed grazing systems incorporate the use of the appropriate grazing animal (in relation to the target vegetation) and the correct stocking rate for the carrying capacity of the target area.

In general, sheep prefer forbs, goats prefer woody vegetation, and cattle prefer grasses. Traditionally, goats and sheep are the primary choice to manage rangeland and forest sites. For example, goats and sheep have been used to control invasive plant species such as yellow star thistle (*Centaurea solstitialis*) throughout rangelands in California. However, cattle are often utilized in grasslands that contain more herbaceous plants, such as forbs. Managers must remember to apply prescribed grazing at the appropriate time of the year to target specific vegetation when plant parts are most palatable for grazers. The duration of the grazing period is also determined by the targeted plant community. To reduce the risk of weed/invasive plant seed from spreading during prescribed grazing, livestock should be given supplemental feed that does not contain weed forage for up to at least five days prior to prescribed grazing activities.

Determining stocking rates is essential when correctly and effectively employing grazing as a management tool. Managers need to know how much forage an animal or group of animals will consume and how much forage is available to them in the grazing site in order to set the appropriate stocking rate (carrying capacity).

The *animal unit month* (AUM) concept is the most commonly used method to determine carrying capacity of grazing animals on rangelands. The AUM provides the approximate amount of forage a 454-kilogram (1,000-lb) cow with a calf will consume in one month. The AUM was established to be approximately 363 kilograms (800 lb) of dry weight forage (not green weight) for a cow with a calf. For animals other than livestock, forage consumptions are calculated using *animal unit equivalents* (AUE). An AUE is the estimated forage consumption of an individual of another species as a percentage of the cow- and calf-based

AUM. For instance, a Spanish goat (*Capra aegagrus hircus*) has an AUE of 0.16, which indicates it eats 16% of the monthly forage of a cow.

Planning

Planning and managing wildlife grazing habitats require continual assessment of vegetative communities and a thorough understanding of grazing behavior and the animal's dietary needs. Understanding grazing animals' dietary and water needs and landscape habitat preferences helps determine grazing behavior and ecological impact of herbivory. For example, white-tailed and mule deer prefer woody habitats, browse, and mast, whereas elk frequent open areas and prefer herbaceous vegetation (Rickel 2005). Understanding plant communities and herbivory reduces wildlife and livestock competition; cattle and elk, for example, are more likely to compete since they make more similar dietary choices than do cattle and deer (Treadaway et al. 1997). However, competition rates fluctuate throughout the year because of density and plant diversity changes with the season.

Even though the list in table 4.6 is not comprehensive, it does provide an overview of how wildlife dietary needs and forage consumption overlap. The dietary intake of grasses, forbs, shrubs, and annuals varies by animal species and biome. Designing a grazing management plan for wildlife and livestock should take into account distribution of vegetation communities and water supplies and competition for space by livestock and/or between wildlife so that habitat factors can be met to sustain wildlife habitats. Hence, land managers must determine and evaluate wildlife habitat needs and plan accordingly. Table 4.7 provides a list of habitat factors that must be considered when constructing a grazing plan. These factors are essential to wildlife sustainability and must be assessed and evaluated on a regular basis throughout the grazing phases because a shift in one habitat area can cause stressors in another, thereby reducing wildlife habitat quality and affecting individual or wildlife populations.

Grazing management is dependent on species, location, and objectives. The overall goal is to maintain or improve range condition as the needs of grazers are met. Managers must consider a host of temporal issues such as growing seasons, reproductive seasons, breeding seasons, wet/dry seasons, and environmental stochasticity (randomness). Proper use of grazing can benefit rangelands, but mistakes can seriously degrade ecosystems. One common issue occurs when grasses are grazed too early in the growing season, thus degrading the area. Grasses that are allowed to grow longer are often better able to sustain

Table 4.6. Dietary overlap for forage classes among bison, elk, white-tailed deer, and mule deer

| Species | Location | Biome | % Dietary Overlap | | | |
			Grasses	Forbs	Shrubs	Annual
Bison	Wichita Mountain National Wildlife Refuge, OK	Grassland	1.0	1.0		2.0
	Pawnee Grasslands, CO	Grassland	41.0	3.0	0.0	44.0
	National Bison Range, MT	Grassland	2.8	1.4	0.1	4.3
Elk	Wichita Mountains National Wildlife Refuge, OK	Grassland	0.1	24.0	0.0	24.1
	Pawnee Grasslands, CO	Grassland	2.9	4.6	3.8	11.3
	Trickle Mountain, CO	Shrub-steppe	3.5	11.5	45.5	60.5
White-tailed deer	Wichita Mountains National Wildlife Refuge, OK	Grassland	0.0	99.0	0.0	99.0
	National Bison Range, MT	Grassland	2.9	24.8	18.5	46.2
Mule deer	Yellowstone National Park, WY	Shrub-steppe	16.6	9.2	52.0	77.8
	National Bison Range, MT	Shrub-steppe	2.3	27.1	33.3	62.7
	Northeastern California, northwestern Nevada	Grassland	2.3	4.2	82.3	88.8
	Trickle Mountain, CO	Shrub-steppe	3.5	10.2	79.5	93.2
	Sheldon National Wildlife Refuge, NV	Shrub-steppe	5.0	31.0	39.0	75.0

Source: Rickel (2005).

Table 4.7. Habitat factors that must be considered when constructing a grazing plan

Edge effects, interspersion of habitats	Animal diversity and grazing seasons
Permanent food and cover	Proper control of plant structure and succession
Carrying capacities	Use of plantings
Fencerow developments	Travel lanes to connect habitats (both on and off the farm or ranch)
Use of native plants	
Water quantity (permanent and temporary sources)	Other appropriate factors, such as nesting, fawning, and calving sites
Water quality	

Source: Natural Resources Conservation Service (2003).

moderate levels of grazing. Proper timing allows a manager to use grazing as an effective management tool.

Effective and efficient grazing management plans incorporate forage and water supplies (water supplies are addressed later in the chapter). To minimize wildlife and livestock competition for forage, the grazing plan needs to identify and assess the plant communities within the context of the animal's grazing behavior. Initiating and integrating herbivory grazing patterns allow for lands to be better managed and reduce overgrazing. Also, understanding and incorporating Holechek's (1980) grazing forage allocation principles listed in table 4.8 provide land managers with grazing patterns, animal distributions, and plant community trends. These factors can either initiate or intensify dietary overlap within and between wildlife and livestock.

Balancing wildlife and livestock rates and population densities is a common management tool to reduce dietary overlap issues. Acquiring private lands through leasing or purchase expands grazing areas, especially during the winter or periods of low forage production. Monitoring weather conditions can help determine amounts and quality of vegetation to expect in the area of interest, which can greatly assist planning and help decrease dietary overlap. Sometimes environmental conditions, especially if lands are degraded, will call for reduction or removal of wildlife or livestock.

Figure 4.8 provides an overview of pasture designs in rangeland pasture primarily consisting of brush species that can provide wildlife with habitat and edge cover. However, the concept displayed can also be duplicated in other types of pastures and croplands. Nonetheless, the most favorable design option will be determined by the type

Table 4.8. Grazing pattern overview

Animals with broad food habits are more able to endure low forage conditions.

Larger ruminants are more able than smaller ruminants to modify diet selection.

Less preferred areas may be selected during or after periods of severe disturbance.

Animals may be more affected by forage availability prior to than during critical periods.

Habitat use may change depending on wild/domestic animal numbers.

Selected vegetation communities may decline in abundance if grazed by only one type of herbivore species.

Source: Holechek (1980).

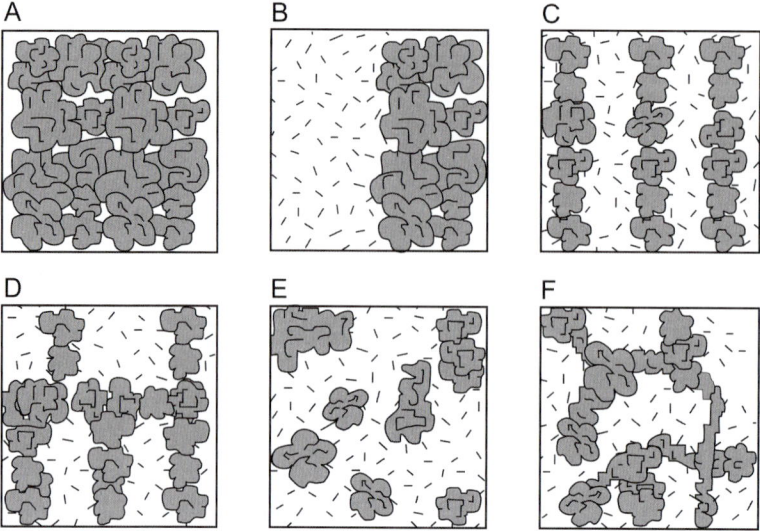

Figure 4.8. Overview of pasture designs. From NRCS (2003, example 8.1).

of animal species, plant community, and plant diversity and intensity. Land managers also need to determine if interactions between animals or species can be described as neutral, mutual, competition, protocooperation, commensalism (one population benefits, but others are not affected), amensalism (one population is inhibited, but the others are not affected), or parasitism and predation (one population depends on the other or directly impacts the other; Natural Resources Conservation Service 2003).

Costs

Good record keeping and correct cost calculations are important components of a wildlife and habitat management plan. By estimating treatment (grazing) costs correctly, a manager can evaluate the cost benefits (value of the treatment's return). Grazing costs, however, are complex and difficult to calculate. Cost factors vary and correlate to pasture/grazing site configuration, labor, managerial ability, and personal preference. Grazing land management systems pivot on the short- and long-term trade-off between projected benefits, carrying capacity, stocking rates, and other cost factors (Conner, Williams, and Dietrich 2003). Cost factors traditionally include location of site, location of leasing operator, location accessibility, availability of water, shelter cover, topography, climatic conditions, and size of target area. Also, the variation in specific costs needs to be included in grazing economic plans.

Economic trade-off also needs to be evaluated between livestock production and wildlife hunting and leasing revenue. The cost of grazing will increase if supplemental feeding of hay or alternative feed (grain, soybean feeds, etc.) is utilized during drought or winter seasons. These costs can also increase because of other factors, such as the distance supplies must be transported. Leasing livestock, goats and sheep in particular, can vary from area to area and can range from $20 to $200 per acre depending on availability and demand.

Implementation

Implementation of grazing systems for livestock and wildlife requires palatable vegetation to be available. Grazing systems should be viewed and implemented as land management tools that contribute to good range condition and sustainability of natural resources. Understanding grazing dynamics facilitates the implementation of successful grazing management plans because stocking rates, forage quality, terrain, and climatic conditions play key roles in best management practices.

Implementing wildlife grazing systems also requires an understanding of herbivory between animal species and plant response closely tracked by observing plant biomass, production, and composition (Fernandez-Giménez and Swift 2003). In addition, it requires setting aside grazing reserves and developing grazing plans that are flexible, manageable, and affordable. Proper stocking rates not only reduce the risk of overgrazing but also minimize environmental degradation caused by erosion or sedimentation. Furthermore, grazing management goals should be assessed, monitored, and modified as necessary as wildlife grazing needs change throughout the different animal grazing phases.

Monitoring and Assessment

Monitoring, evaluating, and assessing grazing activities of wildlife and livestock reduce competition and allow for their sustainable coexistence. If grazing management goals are not being achieved, then the goals and objectives should be reviewed and revised accordingly. Management practices should be modified as necessary to ensure that both short- and long-term goals are maintained. Wildlife habitat needs and food sources should also be monitored on a regular, continuous basis to minimize negative effects on their density and population growth. Supplemental feeding or food plots should be considered as a measurement of BMPs and be implemented during food shortages or periods of drought or low forage production. Livestock and associated support activities (e.g., water, supplemental grain) can also attract undesirable animal species that can negatively impact desired wildlife species.

Management actions must include plans for dealing with these issues (e.g., brown-headed cowbirds [*Molothrus ater*]).

Monitoring species (both livestock and wildlife) population and carrying capacity is also critical throughout the grazing phases. Failure to properly and accurately monitor wildlife population growth can be detrimental to a grazing site; it can potentially take years to restore degraded areas. Developing a systematic method to quantify density, distribution of wildlife populations, habitat, and plant diversity, along with land improvements throughout the year, will make it easier to determine species abundance, distribution, and occurrence (Krueger and Dillard 2013). Maintaining records of these different elements can help managers identify treatment weaknesses and strengths, allowing them to make adjustments to future treatment plans. Besides offering beneficial information to managers, record keeping is also required for tax purposes and open-space tax valuation.

Harvest Management

In this section we discuss the relationship between wildlife harvesting (Aldo Leopold's "gun") and habitat management. Wildlife harvest is the regulated take of wildlife biomass. We focus on the impact that changing habitat has on wildlife populations more often than on the impact of populations on habitat for several reasons: (1) Directly managing wildlife populations is harder than altering other components of ecosystems. It is difficult to force many wildlife populations to change composition or behavior directly. More commonly we try to understand the wildlife-habitat relationship and alter the habitat to effect a desired change in the wildlife. (2) Habitat changes through harvest management are often limited to locations with species in the hunter-favored orders, Artiodactyla (even-toed ungulates such as deer species) and Carnivora (omnivores and carnivores such as bears and big cats). (3) We already manage these systems as we change, destroy, fragment, and sometimes restore habitat. Despite these realities, there are important circumstances where we manage wildlife populations directly, for example, game species (species harvested by humans), species of concern like those that are threatened (potentially will become endangered without action) and endangered (likely to go extinct without action), or wildlife that impacts humans culturally or psychologically (e.g., charismatic megafauna, large animal species that have popular appeal).

Native wildlife harvest can be considered another habitat management tool because there is a link between wildlife population size and composition and habitat characteristics and quality. Wildlife population size impacts energy offtake (e.g., foraging) and energy availability on the landscape. This can then impact other wildlife species sharing the same landscape, such as white-tailed deer (Decker and Connelly 1989; Horsley, Stout, and deCalesta 2003).

The popularity of many game species, such as white-tailed deer, as a target for hunters provides managers with an opportunity to manage habitat indirectly through adjustments in the target population. Successful harvest strategies must integrate human social preferences, accurate species and population information, and reliable population models. Currently, wildlife harvesting (hunting) contributes to population management and individual and population health and provides funding for conservation efforts.

Function and Benefits

Harvest management of native US species is an effective tool for population and habitat management when used properly. It is important to remember that management that incorporates harvesting must combine ecological realities (e.g., goals, harvest quotas, habitat quality, species ecology) with needs and desires of the hunting and nonhunting communities. Effectively run harvest programs therefore incorporate both stakeholder input and ecology. The sustainability of hunting is dependent on the "supply" of target species and demand of the hunters and varies from year to year (Robinson and Bennett 2004). Though not always true, harvesting is often dependent on hunter interest in the problem species (e.g., beavers [*Castor canadensis*] are an important species that could sometimes be controlled by hunting but are less popular hunting targets). If managers identify a candidate for harvest-based management, then they must create or increase hunter interest in this species (e.g., feral hogs [*Sus scrofa*], nutria [*Myocaster coypus*]) or rely on traditionally popular species such as white-tailed deer. Simply put, not every species or population is appropriate for harvesting.

Properly conducted harvest management provides a useful tool for population and habitat protection. Populations at or near carrying capacity can be reduced to prevent damage to or allow recovery of habitat. Managers can regulate hunting to alter population demographics for healthier long-term relationships between populations and habitat. For instance, hunter preference for bull elk may reduce the number of males and cause females to breed in second or third estrous cycles, thereby calving later in the season and lowering offspring survival (Carpenter 2000). Proper management can reduce or eliminate this problem. Peterson et al. (2010) argued that hunting can link humans with nature in a way often prevented by modern society. Joined by other nonconsumptive activities, hunting exposes humans to basic aspects of nature and potentially aids in societal efforts to conserve ecosystems. Finally, some commercial interests are impacted by harvestable species and might benefit from concerted hunting strategies. For example, black bears (*Ursus americanus*) can damage merchantable timber trees in northern California, and feral hogs damage crops and rangelands throughout many US states. Reductions in target populations or changes in target behavior may benefit commercial interests.

Application

Harvest management success combines a coherent plan and reasonable goals, which might focus on healthy wild populations or maximum production. A number of components influence these goals, including social aspects like hunter preferences, ecosystem aesthetics, and economics (e.g., black bear damage to harvestable trees). The objectives to meet established goals might include control of population size, adjustment of demographics, maintenance of individual and population health, and protection or manipulation of habitat.

Every manager with a hunted population must balance populations with habitat integrity. Wildlife harvest limits are often set at levels that maintain a sustained yield (allow offtake without reducing the core population year to year). Sustained yield can be divided into two broad philosophies: (1) maximum sustained yield and (2) optimum sustained yield.

Maximum sustained yield is the highest number of individuals that can be removed without reducing the core population. The population does not increase in this scenario because all "extra" individuals are removed. This philosophy has a variety of shortcomings. Hunting quotas are based on population estimates with intrinsic error. Even small overestimates of population can lead to overhunting, population declines, sex ratio and age structure changes, individual and population health declines, and increased hybridization (Carpenter 2000). This also maintains the smallest self-sustaining population. which leads to increased hunting difficulty and scrutiny from other resource users, such as nature photographers.

Optimum sustained yield incorporates social and biological components into formulation of hunting quotas. The optimum sustained yield strategy maintains higher populations and allows fewer harvested individuals. Higher populations are closer to habitat carrying capacity and generally have lower reproductive output. This means there are fewer "extra" animals to be harvested each year because fewer young are available to replace those removed by hunters. These higher populations ostensibly provide easier hunting opportunities for permitted hunters, more mature and desirable animals, and increased opportunities for older, mature males to pass on genes.

Some managers also hedge their bets by creating matrices of hunted and unhunted parcels within their managed lands. This is helpful because unhunted parcels provide a refuge for individuals in the target population, thus lowering the severity of overhunting, and a well-researched matrix of hunted and unhunted parcels tends toward conservative harvest (McCullough 1996). This lowers the need for comprehensive population estimation, as harvest trends can help illuminate population status. Additionally, management of hunted but difficult-to-study populations (wildlife in rugged terrain, elusive species, less available funds for research and management) can benefit from designated refuges and lowered enumeration requirements. This system requires that managers know important population parameters, such as movement between managed parcels and recruitment (addition to the population) within parcels.

Planning

Hunting regulations and associated management strategies are often based on population growth models. Hunting regulations in the United States are state controlled and created through a process of managing agency, public, and wildlife commission review. Biologist and manager input informs modelers, who use data about fecundity, population size, survival, and past harvest rates to make educated guesses about population sizes (Carpenter 2000).

The application of harvest management is situational and varied. Planning must include decisions about allowed hunting tools (e.g., bow, crossbow, rifle, muzzleloader), method (e.g., baiting, stalking), dates (open and close of hunting season), focal species, focal sex and age class, spatial structure, and management goals. For instance, populations above carrying capacity may qualify for population reductions to protect vegetation and water quality in a habitat, individual and population health, or human interests (e.g., to avoid deer-vehicle collisions).

Harvest management strategies are designed to meet population objectives. Poorly planned targeted hunts can cause unbalanced sex ratios and selection against hunter-favored animals, such as large, antlered males. Managers must carefully monitor populations before, during, and after targeted hunts. Harvest management must begin by setting goals for the population of interest that includes some determination of preferred population size best for both ecosystem health and hunter satisfaction. This requires accurate estimates of population size, health, and demographics generated by robust techniques, such as mark-recapture and distance-based efforts. These data can then be used to generate predictive models for the target populations. Managers use these models to determine appropriate hunting quotas for the next hunting season. These hunts can generally be manipulated by limiting licenses, for example, through a lottery system. Managers can then monitor the results of harvests through hunter check stations, mandatory reports associated with licenses, mail surveys, and telephone surveys (Carpenter 2000).

Increasingly, hunting on private lands is an important component of harvest management strategies. Private landowners can form land management plans with state and federal agencies that include a hunting component. Private landowners charge an entrance fee for hunters in addition to other possible fees, such as lodge cost or guide fees. Hunters benefit from less hunting competition, and landowners benefit from the additional income. Unfortunately, wildlife managers have difficulty monitoring the harvests on private lands without set agreements in place. Planning for future harvests can be impeded by subsequent inadequate data.

Social considerations now must be considered in harvest management planning. Humans have vested interests in wildlife management decisions because wildlife populations have the capability to impact human health and economics (e.g., property damage, nonconsumptive benefits to society). Humans also tolerate harvest strategies differently based on location (rural vs. urban/suburban) or situation (human-wildlife conflict present, cultural history of hunting). The manager must evaluate the tolerance for wildlife and wildlife management by communicating with stakeholders.

It is difficult to provide set timing for harvesting since all wildlife species are unique. The timing of harvest is determined by ecological and social influences that vary greatly. Considerations of juvenile independence, wildlife stress, interspecies links (e.g., predator-prey dynamics), and habitat quality are important when determining the timing of regulated hunts. Carpenter (2000) argued that longer hunting seasons help support the hunting tradition and ease the hunter crowding inherent in shorter seasons. Carpenter also argued for more micromanagement of hunting seasons. For instance, scheduling opening day early in the week spreads the hunting pressure evenly throughout the week.

Harvest management must include a cogent plan and clear goals with detailed understanding of the current and future population, demographics, local ecosystem, animal ecology, disease concerns, and urban/suburban impacts. Harvest strategies must be based on sound ecological data, including breeding cycles, genetics, and independence of young during the hunt, and impact of hunters on animal behavior. Misunderstanding or ignorance of these issues can cause increased indirect adult and juvenile mortality, lowered fecundity, and other issues beyond direct hunter-caused mortality. Managers must determine the minimum viable population, number of successful breeding individuals (effective population size), and connectivity among populations. These are critical concerns for long-term population persistence and genetic health.

Management plans must change according to new parameters because environmental and demographic stochasticity can impact populations and habitats season to season. The increase in hunts on private lands, which are difficult to monitor, adds to the complexity. Such hunts also raise questions about management of game species as a public resource and private lands management strategies (e.g., high fencing).

Not all wildlife or situations are equal. Some large herbivore populations, such as white-tailed deer, are more easily and more accurately estimated than elusive and less numerous carnivore species such as mountain lions (*Puma concolor*). Population estimates are generally less precise in areas that are difficult to survey, such as rough terrain, mountain ranges, aquatic environments, or restricted areas. Some areas that are actively managed for natural resources extraction such as timber or mining might experience serious fluctuations in habitat that impact population structure of target species. Additionally, species with low reproductive output or highly spatially structured populations (metapopulations) are poor candidates for sustainable harvesting. Harvest of these species may cause subpopulation extirpations and serious population declines, as hunting often disrupts dispersal and recruitment (McCullough 1996).

Management directions are dependent on a host of considerations besides hunter approval (Peterson 2004). Management strategies are increasingly scrutinized by an involved public with varying concerns. Human social considerations are widespread and increasing in importance, including issues like aboriginal hunting rights, animal rights, social legitimacy of hunting, and commercialization of wildlife. Managers have multiple stakeholders that must be involved and their unique perspectives addressed.

Costs

Harvest management costs are affected by species, season, hunting methodology, social aspects, location, and management personnel required. Regulated hunting can provide managers an economical method of population management, but the costs are difficult to calculate. The basic costs to hunters are license fees (determined by state), travel costs, and associated equipment. Managers must consider infrastructure, equipment, and employee costs when implementing harvest management strategies.

Implementation

Harvest management requires species for hunting, appropriate habitat, a receptive public, and clear, achievable management goals for wildlife and habitat. Harvests are a powerful management tool with both positive and negative potential. Managers must have a clear understanding of the relationship between the hunted species and habitat. This relationship is critical for population sustainability and habitat conservation. Successful harvest programs also require accurate and timely data about target populations and associated species and habitats. These data must be interpreted correctly with robust models and common-sense evaluations. Also, hunting strategies require a substantial investment of time and effort by managers. This may require advertising new but impactful species (species that have an important impact on the ecosystem) to hunters, involving the public in hunting policy decisions, gathering critical data, and making difficult decisions about hunting seasons and quotas. Finally, data collection must continue during and after the hunting seasons to ensure management effectiveness and inform future strategies.

Monitoring and Assessment

Monitoring and assessment of harvest management programs are critical for planning and problem solving. This adaptive framework requires detailed data collection of hunter take, population size and demographics, habitat characteristics, and seasonal fluctuations in weather and climate. Determination of the success of a program dictates future directions and provides managers with information to provide to the increasingly involved public.

Monitoring and assessment are often overlooked to the detriment of management programs. The temptation to thoroughly research harvest programs prior to implementation but neglect monitoring and assessment is driven by several realities. Managers are often charged with managing at landscape or ecosystem scales with variable funding and a fluid workforce. The impetus and funds are not always available to adequately monitor all programs. The complexity of natural systems requires this investment to most effectively utilize management programs. For instance, Kilgo, Labisky, and Fritzen (1998) found that hunting white-tailed deer in Florida impacted deer population dynamics and behavior enough to potentially impact Florida panthers (*P. c. coryi*). Although it is somewhat intuitive that prey population fluctuations would impact predators, the exploration of this interaction required expanded research and consistent monitoring and assessment of the white-tailed deer harvest program.

Herbicide Applications

The chemical application of herbicides is another land management tool (Aldo Leopold's "plow") that can be used to control vegetation in rangelands, forests, and aquatic ecosystems. Herbicides are used to control, manage, or suppress vegetative growth and/or establishment of undesirable plant communities. Chemical application methods vary and are dependent on time of year, the growth or establishment stage of vegetation, moisture conditions, proximity to water (ground and surface), and availability and training of equipment and operators. For example, cacti management is very time-of-year-specific due to the resprouting properties and drought tolerance of cacti. Herbicides can be plant derived or synthetically manufactured; however, most brush management herbicides are chemically engineered. Nonetheless, both types of herbicides require safety measures when handled, transported, or applied. The target species' relation to the presence of desirable vegetation, soil properties, and water proximity determines the type of herbicides or chemicals that are best and most effective on specific target species (Bussan and Dyer 1999).

Chemical applications are most effective in controlling or managing stands of single species in areas where desirable plant communities are scarce or absent. Also, the use of herbicides in remote or difficult-to-reach areas is a common practice due to the difficulty of transporting heavy equipment and personnel. In addition, rhizomatous species of unpalatable plants often require repeated treatments and/or physical removal for proper management and control. All non-domestic use of herbicides needs to be applied in accordance with the law and the manufacturer's labeled recommendations (proper equipment, safety measures, and training). The commercial and agricultural application, vending, licensing, and training are regulated by the federal government and subject to penalties and fines.

Function and Benefits

Herbicides are safe and effective when properly applied and can increase property value and livestock productivity and enhance the aesthetics of rangelands, forests, and aquatic areas. Integrated management plans should include alternative vegetation control methods that can be combined with or without the use of herbicides. No single herbicide method, application, or rate can successfully and systematically manage undesirable vegetation. Chemical management plans also need to assess the potential risks to humans, water sources, and animals as well as the advantages and disadvantages of the various associated methods and cost factors.

The most common uses of herbicides are to (1) increase forage production and forage quality; (2) improve herbaceous composition; (3) manage, control, or evade invasive plants; (4) suppress brush; (5) improve aquatic habitat and protect water quality; (6) increase aquatic biodiversity and recreational use; (7) improve livestock and wildlife habitat; (8) promote tree growth; (9) control parasites and pests; and (10) improve land accessibility and visibility. The control and mortality rates of herbicides range from low (0%–35%), high (56%–75%), and very high (76%–100%) and are obtained when treatment is properly applied under the optimum conditions.

The *mode of action* is the mechanism or interaction process by which the herbicide kills or causes damage to the plant. Mode of action processes involve (1) contact and absorptive action of herbicides into the plant via cuticle or epidermal root tissue, (2) translocation or movement of chemical(s) via the cellular level, and (3) mechanism of action that causes biophysical or biochemical damage. Understanding the mode of action and herbicide chemical properties can minimize damage or injury to nontarget plants and prevent herbicide resistance problems. The herbicide chemical properties are provided on labels and material safety data sheets (MSDSs). The active ingredients identify the chemicals that cause damage to vegetation (and potentially to the environment). The other ingredients are chemicals that improve application (adjuvant) or suspend herbicides during application (diluent, solvent, or carrier). Since herbicides are either species selective or nonselective (broad spectrum), their fate after application needs to be carefully considered, especially when selecting broad-spectrum herbicides, because they can cause damage to vegetation and ecosystems long after application. Furthermore, equipment has to be properly calibrated and adjusted to adequately control application volume and concentration.

Poor equipment calibration can lead to illegal application rates and cause environmental toxicity.

Application

Many chemical applications and combinations of applications exist that have both advantages and disadvantages. However, special consideration needs to be given to climatic conditions because they determine plant uptake and adjuvants that enhance application/herbicide activity for optimal plant uptake. Temperatures following chemical application are also important and can determine crop injury or safety. Precautions need to be taken when temperatures increase above 29°C (85°F), as some herbicides (e.g., difenzoquat, bentazon, lactofen) become more active. Plant growth stages also need to be considered; and in some cases, it is acceptable to apply reduced rates when target plants have developed beyond the optimal treatment size and when temperatures surpass 32°C (90°F). Furthermore, a thorough understanding of herbicide mechanics not only increases effectiveness but also minimizes nontarget plant injury and herbicide resistance problems.

The most common chemical applications include (1) wick or swipe (manually applied); (2) broadcast (aircraft, tractor, truck, etc.); (3) spray tanks (handheld or backpack sprayers); (4) mountable tanks equipped with spraying mechanisms; (5) foliar (direct application onto leaves of plants); (6) basal (herbicide is mixed with a penetrant oil like diesel and directly applied to the bark of standing trees); (7) hack and squirt (bark is cut in a downward motion, creating cup indentation for the herbicide solution); (8) cut stump (cutting tree down and immediately applying herbicide solution to the trunk surface); and (9) soil spot treatments (treating individual specimens by applying chemical directly to the surrounding soil). Table 4.9 provides a list of common herbicides.

Planning

Before initiating chemical treatment, managers need to develop a comprehensive land management plan that includes other land management options (or a combination of methods) to obtain the desired goals and objectives. Herbicide treatment goals and objectives should be properly and accurately identified to obtain desired outcomes. Comprehensive and thorough herbicide treatment plans minimize risk factors and prevent catastrophic outcomes. The plan also needs to determine whether herbicide treatment is the most appropriate and cost-effective method and if the projected time frame is expected to

Table 4.9. Commonly used herbicides

Aquatic herbicides

Copper sulfate (or blue stone): for algal treatments

GreenClean, PAK 27, Phycomycin: controls blue-green algae

Habitat: effective on post-emergent floating and emergent aquatic vegetation

Navigate, Aquakleen, Renovate, Reward, AquaMaster, Eraser AQ, Touchdown Pro, and AquaNeat: absorbed and moves within the plant to the site of action

Range/forest herbicides

Clopyralid (Pyramid R&P, Reclaim, Transline, Stinger, Confront): controls unwanted plants on ranges, pastures, rights-of-way, lawns, and turf

Picloram (Picloram 22K,Tordon 22K, Grazon): controls woody plants and broad-leaved weeds; however, most grasses are resistant

2, 4-D: varies in solubility/volatility and used to control many types of broad-leaved weeds

Tebuthiuron 20%/80% (Spike 20P, Spike 80DF): controls woody plants and weeds in noncrop locations

Glyphosate (Roundup, Rodeo, Pondmaster): perennial plants (grasses, sedges), broad-leaved weeds, and woody plants

achieve goals and objectives. If chemical treatment is found to be the optimal management tool, then safety precautions need to be incorporated into the herbicide treatment plan. Managers also need to have a thorough understanding of the risks of chemical exposure to and effect on plant communities, wildlife, livestock, bodies of water, the environment, and humans.

The treatment plan must identify and describe all the potential hazards (e.g., local hydrology, sensitivity to soil compaction) found near the targeted area. It also needs to identify short- and long-term environmental goals; weather conditions projected for the treatment date; a list of equipment, herbicides, and supplies; and the type of training required to safely conduct the chemical treatment (Scifres and Hamilton 1993).

Because timing and temperature conditions during treatment determine the effect on both target and nontarget plant species, chemical application management plans need to incorporate forecasted weather conditions before and after application. Herbicides are more effective when applied at temperatures that are between 18°C (65°F) and 29°C (85°F) (North Dakota State University Extension Service 2013). During cold or freezing temperatures herbicide uptake and translocation slow down and can potentially result in a lower appropriate lethal dose reaching the target vegetation. Low humidity levels and soil factors such as moisture level and amount of organic matter can also play a role in herbicide uptake and effect. For example, low humidity and soil water can result in growth of stressed plants, which often have smaller stems with thicker cuticles and more epicuticular wax that retard retention and penetration of herbicides (Lundkvist 1997). Hence, weather conditions that favor optimal plant growth will also be optimal for herbicide application.

The environmental checklist in table 4.10 is provided to assist aquatic and land managers during the chemical application planning process. This checklist is not intended to be the sole source of best management practice considerations and therefore should be used only as a guide.

Costs

Herbicide costs vary and depend greatly on the herbicide management plan, as the target area, methods of application, and plan goals and objectives will require particular chemical application and equipment. Although it is difficult to calculate actual cost per acre at any given time due to price fluctuations, rental equipment costs, and so forth, it

Table 4.10. Summary checklist for herbicide application

Chemical information

Obtain all safety information including the Material Safety Data Sheets (MSDSs), provide safety directions, and provide appropriate handling of chemical training.

Obtain chemical application and transportation requirements (including state and federal regulations).

Obtain information on storage of chemical requirements and conditions.

Chemical application

Make sure that anyone handling chemicals has appropriate and up-to-date training in accordance with the 1999 Pesticide Act regulations.

Place appropriate spill or cleanup kit in close proximity to chemical storage and when transporting, mixing, or applying.

Place first-aid kit and wash water in close proximity to chemical storage and when transporting, mixing, or applying.

Store or mix chemicals away from bodies of water, high-water tables, and areas where runoff may occur.

Post and strongly enforce post-spray reentry time periods and make sure to provide notification.

Complete chemical application records in accordance with the 1999 Pesticide Act regulations.

Mix and apply chemicals as specified on the product label.

Contain spray drift and identify, monitor, and manage risks to sensitive areas during and after chemical application.

Source: Monsanto (2009); Agriculture Victoria (2016).

is important to seek current prices for required items to determine an actual projected cost. Applicator contracted services and equipment rental are sometimes the most expensive costs associated with herbicide application. Expenses depend on the size of the area being treated, equipment rented or purchased, chemical cost, travel expenses, daily rate of trained personnel, and number of personnel necessary to implement the herbicide treatment.

Implementation

Before implementing an herbicide management program, the manager must first determine if herbicide treatment is the most effective in terms of cost and vegetation control or manipulation. If so, which chemicals and treatment applications that will yield the projected land management goals are then identified. The manager should also consider a combination of treatment methods (e.g., prescribed fire and herbicide, mechanical and herbicide treatments). Combining land management treatments often reduces application costs, extends results, and minimizes risks of chemical exposure.

Therefore, herbicide treatment options should be cautiously assessed, and the benefits should always outweigh the risks, exposure, and costs. Chemical site management plans should be developed and revised as necessary to obtain the desired projected short- and long-term goals. Recall that nonresidential application of herbicides must be applied by trained personnel. Chemical application has to be conducted in accordance with the law and land management goals. Safety is always the number-one goal, and chemical application is always carried out under specified weather conditions along with properly calibrated application equipment. The site location, acreage, target vegetation, and weather conditions will determine the precise course of action; application method is selected according to the strategic goals set forth by the land manager in conjunction with federal, state, and local agencies. Understanding federal, state, and local laws with regard to transportation, storage, and application of herbicides is also essential, as the transportation and use of certain chemicals are prohibited in some states.

Monitoring and Assessment

Monitoring, evaluating, and assessing herbicide management plans as well as understanding chemical exposure reduce risk factors. Environmental processes and translocation of herbicides minimize environmental pollution and protect aquatic habitats and water quality. If for some reason herbicide management goals are not achievable, then the plan should be amended to achieve the desired goals.

Effective monitoring and herbicide treatment assessment provide the basis for objective evaluation; whenever necessary, modifications are made to the management plan to ensure that land management goals are achievable and implementable. The land manager needs to assess the ecological impact to determine if the project objectives will be met. Reviewing and outlining again the course of action before herbicide implementation can minimize application error. Also, precautions must be taken when the temperature increases above 29°C (85°F) because some chemicals become more active and potentially more dangerous.

Supplemental and Replacement Water

Provision of supplemental and replacement water is designed to increase water availability in areas of interest. Many regions now suffer from declining or exhausted water supplies due to agricultural and urban uses. Strains on surface and subsurface supplies, coupled with topographical and hydrological changes, result in limited water for domestic and wild animals. Supplemental water and replacement water are critical in maintaining domestic and wild animals on the landscape. Managing wildlife may, therefore, require restoring, redistributing, or redesigning water sources to increase water availability. Sometimes restoration requires protecting water supplies from domestic livestock. Other times, this might mean repairing marshes, wetlands, ponds, or creeks. This strategy can be more economical than projects starting from scratch, such as drilling water wells, because restoration projects have an extant foundation on which to build. Redistributing or redesigning water systems often requires managers to evaluate problems and create solutions (e.g., reduction or elimination of a well, trough and windmill overflow).

Both wildlife and livestock need to have access to fresh, clean water at all times to remain productive and in good health. Supplemental or replacement water systems geared toward livestock can also be utilized by wildlife and should be incorporated into grazing systems because availability and accessibility of fresh water increase animal energy efficiency and performance. Water is sometimes provided during drought seasons; however, it is most often provided in areas where water is limited or not accessible. Unfortunately, many individuals mistakenly believe that water sources suitable for livestock are also appropriate for wildlife. This is a misconception, as most supplemental or replacement water sources are unsuitable for young wildlife and many bird species (Texas Parks and Wildlife Department 2007). Also, it is generally not recommended to place supplemental water systems in shaded areas because they often attract loitering and loafing animals such as feral hogs, thus concentrating waste, decreasing water quality, and exacerbating erosion.

Function and Benefits

Wildlife and livestock need to have access to fresh, clean water on a daily basis. An adequate daily supply of good-quality water increases animal energy efficiency and performance, supports animals' general health, and reduces stress by minimizing the areas they enter to obtain water. Human demand for water has reduced or eliminated many water sources for wildlife and, consequently, negatively impacted these benefits for wildlife. Furnishing supplemental and replacement water is often a response to these human impacts on wildlife habitat (e.g., water diversion, creation of impermeable surfaces, introduction of invasive-exotic plants). Projects designed to increase or protect water sources can have large impacts on conservation and are critical components of wildlife persistence. Therefore, well-designed water improvement projects can positively impact ecosystems and a variety of species. For instance, furnishing supplemental and replacement water not only contributes to the establishment and maintenance of wildlife habitat but can also improve wetland functions (water quality and quantity). Restored wetlands can improve water quality and biodiversity and fulfill human preferences (e.g., waterfowl hunting, wildlife photography).

Supplemental water sources improve wildlife habitat utilization and thereby can improve land value. Semiarid rangelands are traditionally poor water producers due to *evapotranspiration* (transfer of water from the earth [evaporation] and plants [transpiration] to the atmosphere), which normally exceeds precipitation (Hibbert 1983), and require supplemental water. However, the restoration of natural water sources can do more than increase water supplies to wildlife. It can also alter the hydrology since restored natural watering systems have the capacity to store large amounts of water and can impact the hydrological cycle of the area.

Application

Well-developed wildlife management plans identify supplemental water particularly during drought seasons or whenever natural water systems are lacking or are in limited supply. Understanding wildlife water needs and available supplies can help determine whether supplemental water sources are needed. In addition, land managers need to understand the difference between adequate and optimal water supplies; water may not be a limiting factor, but additional water facilities in strategic locations might be needed to optimize wildlife habitat for specific wildlife species (Miller 2007). A thorough understanding of supplemental water methods and site selection helps determine the type of water system that could be installed.

Water pipelines, water tanks, 55-gallon barrels, and so forth can be used to collect or distribute water to wildlife habitats. In addition, wildlife water guzzlers can be built to collect rainfall or runoff water and supply water to wildlife. However, these devices need to be part of the overall habitat management program (Texas Comptroller of Public Accounts 2013) and should be strategically placed to supply adequate amounts of water to wildlife. A guzzler, for example, should be placed outside gullies, arroyos, or creeks to avoid flood damage or siltation (Miller 2007) and needs to be placed near food supplies. Placing or positioning tanks with the open end facing north will keep water temperatures down in the summer and minimize evaporation if no shade is available. However, in the winter, the tanks should be positioned facing south to help thaw water (Miller 2007). The appropriate spacing of water facilities is important, as certain wildlife species have maximum travel preferences. Table 4.11 provides optimal and maximum travel distances as well as typical daily use.

Providing supplemental and replacement water entails an array of water options unique to each system. However, identifying water supplementation goals and objectives is key to a wildlife management strategy. Water is essential to all forms of life; however, the need to supplement water on a small or large scale depends on the targeted spe-

Table 4.11. Optimal to maximum wildlife travel distances

Species	Optimum (miles)	Maximum (miles)
Pronghorn	2.0	4.0
Mule deer	0.5	1.0
Elk	0.5	1.0
Chukar	0.5	1.0
Most quail species	0.25	1.0
Pheasant	0.5	1.0
Turkey	1.0	2.0
Mourning dove	3.0	5.0
Songbirds	0.25	0.5

Source: Miller (2007).

cies and availability of other habitat elements such as food, shelter, and cover. Before developing a supplemental watering plan, land managers must conduct a thorough habitat assessment that determines whether water is a limiting factor for the targeted wildlife species. Managers need to understand water-limiting or accessibility factors to properly and effectively assess grazing and watering wildlife plans.

Additionally, the supplemental water benefits of wildlife management need to offset the installation and maintenance costs. Table 4.12 provides a general list of supplemental water considerations. Assessing these guidelines as well as the landscape properties should help wildlife managers determine if furnishing supplemental and replacement water is cost effective and if it will increase wildlife production. The figures and tables listed throughout this section should also be used when planning and evaluating water supplementation options.

Planning

Planning for supplemental and replacement water systems requires a thorough knowledge of landscape and wildlife ecology. Simply placing buckets of water in the desert is unlikely to meet management goals. Managers must understand wildlife water needs and how best to meet those needs. Occasionally, built structures (e.g., rain guzzlers for Sonoran pronghorn [*Antilocapra americana sonoriensis*]) can capture water that the landscape cannot. These must be placed in areas most available to target populations. Managers must also consider wildlife behavior. Water does little good for the population if each guzzler is

Table 4.12. Supplemental water general consideration guidelines

Where livestock or larger wildlife species are present, the facilities should be fenced to provide proper protection.

Plastic and PVC materials can be damaged by rodents and ultraviolet light. As little as possible of this material should be left accessible to rodents or sunlight.

Hard winter freezes can damage some facilities. Provisions should be made to drain or shut off water supply during these periods.

Proper maintenance of equipment will ensure adequate wildlife water and increase life of facilities. As necessary for all equipment, facilities should be checked on a regular basis.

Algae growth can be a problem in many facilities. The less sunlight, the fewer algae growth problems will be encountered. As much as possible, the facility should be shaded. If algae growth becomes too bad, the facility may have to be drained and cleaned.

Source: Turrentine (1992).

guarded by several dominant individuals that prevent access to others. Managers must also consider allowing or restricting access to nontarget species, such as domestic livestock or other wildlife. Water must also be provided at critical times of year, such as fawning months, and in the quantities necessary for population health. Often, managers must decide how to use the landscape to meet water needs (e.g., wetland or river restoration). Managers must then navigate the fractious world of water rights and water shortages since many water problems are human caused.

If water supplementation or replacement is needed, then the process begins with an evaluation of the landscape. What water resources are available? What are the wildlife needs? This is followed by an evaluation of appropriate courses of action and budget. As a general rule, simpler systems are less costly. Additional information on different supplemental water facilities is available through local cooperative extension or wildlife and natural resources government agencies. These agencies may have construction and installation information about systems unique to particular geographic areas. Developing and installing accessible water facilities will increase animal distribution and thus minimize impacts to all water resources. In some cases, it may be necessary to seek technical assistance.

Environmental indicators should be assessed and monitored throughout the water supplemental stages and phases. Also, establishing off-stream water sources can help minimize impacts to wetlands and riparian ecosystems since animals travel to water sources in relationship to food and topography. Improper location or design and construction may increase nonpoint source pollution (precipitation runoff moving pollution from the watershed into the water body). In addition, soil and stream-bank erosion may result. Furthermore, restoration of degraded lands can increase land management cost, and the benefits of supplemental or replacement water may be diminished or never obtained.

Supplemental water and replacement water are ecologically and seasonally dependent. Water systems rely on natural rainfall, surface water, subsurface water, or manual filling by managers. Rainfall-fed guzzlers and wetlands are examples of naturally fed water systems. During periods of rainfall insufficient to meet target species needs, some water systems might be manually filled (e.g., by tanker truck or helicopter) or additional structures added. Occasionally managers permit more water flow through dams during periods of low water availability downstream. All water structures must be regularly and thoroughly monitored to ensure proper function and water availability.

These all need to operate at ecologically critical times and in areas accessible to most of the population.

The major factors to consider in regard to providing supplemental water are (1) plant species (site locations and densities); (2) target species' daily water requirements; (3) soil properties (texture, depth, porosity); (4) soil water-holding capacity (the ratio between depth of precipitation to depth of penetration/percolation measured against the slope of the terrain); (5) space availability for aprons (micro-catchments), tanks, and so forth; and (6) accessibility to the site where supplemental water facilities need to be installed.

Costs

Cost factors associated with supplemental watering are vast and difficult to calculate due to the nature of each system. The cost can be thousands of dollars depending on the type of supplemental water systems chosen. For example, the cost of drilling water wells or a spring can be exorbitant, especially in areas where the water table is deep or accessibility is difficult. Table 4.13 provides a general list of possible costs associated with the installation of supplemental water facilities. This list is not comprehensive but can serve as a general reference to calculate supplemental water costs.

Table 4.13. Supplemental water cost checklist

Restoring natural water sources	**Small-game guzzler**
Construction, land clearing, etc.	Corrugated galvanized metal sheets
Designing contracted services	PVC
Reseeding, revegetation	Post
Soil lining, soil compacting, etc.	2 x 4 lumber
Piping water	Screws
Pipe, pipe fittings, connectors, etc.	Nails
Collecting tank, trough, drum, etc.	**Other materials**
Cementing water collecting area	Fencing material
Fencing area if necessary	Hoses
Float	Floats
Drum with facet or float	Electric pump to pump water
Drum (plastic or metal)	Holding tanks
Facet, float	Soil liners
Metal/wood stand	Soil testing
Hose	Installation cost

Implementation

Supplemental water systems should be viewed and implemented as land management tools to sustain livestock and wildlife production. Understanding the animals' behavior with regard to drinking water will also help determine the ecological impact of the water sites; therefore, they should be properly managed to be sustainable. Managers should also understand grazing dynamics and how land carrying capacity, stocking rates, forage quality, terrain, and climatic conditions play a key role in how wildlife is managed and supplemental water is allocated.

Monitoring and Assessment

Supplemental water facilities can be damaged by animals or environmental elements (e.g., freezing temperatures, metal corrosion) so must be monitored and maintained. Therefore, supplemental water facilities need to be inspected on a regular basis to keep them in good working condition, and natural water sources need to be protected from contamination or livestock. Excluding and/or controlling livestock around springs may also help maintain native vegetation and improve animal diversity. Channeling water from water wells or springs also prevents or minimizes degradation of immediate areas by decreasing trampling.

Periodically, water-drinking locations should be assessed in relationship to food sources and evaluated in relation to the short- and long-term goals. Other wildlife habitat needs also must be evaluated and monitored on a regular basis to maximize wildlife production. Supplemental feeding areas should be consistent with the recommended distance guidelines, especially during stress periods (e.g., drought, low forage production, winter).

Monitoring species population, carrying capacity, and forage availability is also important to wildlife production and water supplementation. Understanding the dynamics of wildlife interactions and functions can help determine the necessity of supplemental water. Failure to properly and accurately monitor supplemental water facilities can result in environmental degradation from overutilization by wildlife or malfunctioning pumping equipment, such as damage caused by water runoff. Supplemental water can also attract predators that prey on the target animal species and facilitate disease transfer by grouping animals together in a limited space.

Urban and Suburban Processes

Urbanization is an area of growing importance in wildlife habitat management. With humanity rapidly changing from a rural to an urban species, urban areas are called on to support an increasing number of species. The size and number of urban areas make them important parts of regional and global biodiversity (Dearborn and Kark 2010) as well as contributors to ecosystem function and services (McClintock 2010). These cityscapes can vary in form and function, from residential to industrial and from high to low population density, but human-built structures dominate the landscape (Anderson et al. 1976). However, these areas still contain vegetated areas, soil and hydrological structures, and buildings that support wildlife. This diversity of spaces can provide winter or breeding habitat for migratory birds, year-round habitat for terrestrial mammals, or occasional refugia for endangered species and invertebrates. Certainly, wildlife is resilient, and we often see waterfowl on water treatment ponds, migratory warblers at backyard bird feeders, raccoons (*Procyon lotor*) in woodlots, native invertebrates such as bees and beetles in gardens, and even coyotes (*Canis latrans*) or mountain lions patrolling the peripheries of suburban areas. Of course, the ability of urban and suburban areas to support animals is dependent on a number of variables, such as the species in question; the number, type, arrangement, and availability of habitat; and the policies of the people dwelling in these urban areas. For instance, mountain lions on the outskirts of a subdivision are most likely an indication of urban invasion into rural or undeveloped areas that supported mountain lions.

For our purposes, when we discuss wildlife habitat management in urban spaces, we generally refer to "green spaces." These can include areas with vegetation, such as urban forests and parks, commercial urban forests, urban agriculture, private yards, riparian corridors, and vacant lots. Urban green spaces can include many vegetation types that support a variety of species (Caula, Hvenegaard, and Marty 2009), but even if urban areas include green spaces, urbanization tends to be a net negative for species diversity.

Human population size generally negatively correlates with species richness across a variety of taxa and situations (Bradley and Altizer 2007). The reasons are both obvious and subtle. Urban areas obviously replace large swathes of native vegetation with highly modified cityscapes that are generally less supportive of native species. Additionally, many of the species in urban areas are exotic species that compete with less urban-adapted native species or are early-successional species better adapted to highly competitive urban areas. Ultimately, urban areas are expanding and generally provide worse locations for native species than their preferred native vegetation types.

However, urbanization has complex, often counterintuitive, impacts on native systems, impacts that are highly dependent on urbanization characteristics and local vegetation types. For instance, avian diversity often increases from urban centers into suburban areas, peaking in rural sites with remnant forests (Clucas et al. 2011). The dramatic land cover changes in urban areas eliminate many native bird species, but suburban areas provide a more heterogeneous landscape that results in increased bird diversity. The loss of urban green space, in this case forest, then promotes replacement of unique niche species with early-successional or commensal species. Although diversity may not strictly decline in this scenario, the native system is dramatically altered as suburban or rural refuges are replaced.

If we dig a bit deeper, we also see less obvious impacts from urbanization, such as changes in wildlife disease ecology or impact zones related to human structures. As an example, research has documented that reductions in host diversity in urban areas increases prevalence of Lyme disease in those species present, such as mice and humans (Bradley and Altizer 2007). Additionally, the impact of development on wildlife habitat is related to distance from the disturbance (Theobold, Miller, and Hobbs 1997). In other words, a house affects the surrounding area by impacting wildlife behavior and changing vegetation structure. If we assume that disturbance declines the farther we move away from a human development, then there are large implications for urban designs. For instance, large lots might seem to provide more open space for wildlife habitat but also spread out negative disturbance associated with those built areas (Theobold, Miller, and Hobbs 1997). This contributes to urban sprawl and the spreading human footprint.

So far we are overlooking an obvious complication for urban wildlife habitat management. The human-wildlife relationship is especially important in urbanization as well. As urban areas expand and consume rural areas, the clash of cultures, ideas, and ways of life among people complicates the already fractious nature of wildlife habitat management. These new realities require concerted efforts to include all stakeholders in wildlife habitat management decision making (Patterson, Montag, and Williams 2003). In fact, wildlife and associated habitat can become proxies in the emerging conflict. For instance, the potential delisting of Yellowstone grizzly bears (*Ursus arctos*) from the federal endangered species list in the first decade of the 2000s created enormous controversy. However, Parker and Feldpausch-Parker (2013) found that a majority of the disagreements were not about grizzlies, even though the name was often prominent in the exchanges, but more about human concerns, such as private property rights, conservation values, cultural changes, and economics. The term "grizzly bear" actually meant a variety of things in this heated environment. Wildlife habitat managers would benefit from treating these disagreements as complex social and ecological problems.

As a final example of the interconnectedness of humans and wildlife habitat in urban settings, we examine urban forestry. We discuss forestry extensively in other chapters and will avoid belaboring the management methods here. Many of the techniques we discuss elsewhere are relevant in urban settings as well. Helms (1998, 193) defined urban forestry as the "art, science and technology of managing trees and forest resources in and around urban community ecosystems for the physiological, sociological, economic, and aesthetic benefits trees provide society." Although only a component of urban wildlife habitat management, urban forestry is important in maintaining areas of forest in cityscapes. These forests can provide many environmental functions and services, such as improved air and water quality, flood control, and pleasant aesthetics (Nowak 2006).

The role of urban forestry is critically important because trees in urban areas represent almost one quarter of all trees in the United States. These urban forests can provide air quality improvements, nature experiences for local residents, flood and erosion control, and a variety of other benefits. However, effective management, such as maintaining tree diversity, promoting larger wood lots, and incorporating silviculture when appropriate, is required to fully realize these benefits (Alvey 2006). For example, urban forests generally include both fragmented native forests and planted trees. As Clark et al. (1997) explain, the native forest remnants may have some limited capability for self-renewal, but planted trees have little to no such capability. Managers must assess the capability and intervene when necessary by planting new trees, managing succession, and mimicking disturbance to continue receiving benefits. The challenge is managing urban forests with a variety of ecological, economic, and aesthetic needs and desires. The many human perspectives often conflict with each other, again emphasizing the need for early and consistent stakeholder involvement, shared decision-making power, and facilitated processes.

Urbanization has a host of negative impacts on wildlife habitat, in fact, far more than we have discussed here. However, many of the management scenarios and ideas we highlight in this book are applicable in these urban settings. For instance, small-scale commercial forestry has a place in many urban forests. Conservation of urban parks and riparian areas has increasing importance to wildlife habitat management, as do hydrological structures, farms, backyards, high-rise ledges, and many others. The multiple challenges relate to any area with high human population density and include creating effective stakeholder involvement, managing areas with numerous landowners, and dealing with city laws and plans not related to wildlife habitat conservation.

Additional challenges are familiar to wildlife habitat managers everywhere and to readers of this book in particular. These include understanding hydrology and soil biochemistry, multiscale ecosystem relationships, energy cycling, and all of the myriad biotic components. Urban systems will only grow in importance as the human population increases. It is critical that humans improve their understanding of wildlife habitat management within these areas.

Stakeholder Involvement

Managers and policy makers are increasingly realizing that stakeholder involvement in decision making is critical to successful management. In fact, many of the conservation and community development endeavors in the United States are becoming more collaborative in nature (e.g., conservation of mountain lion habitat or rural and agricultural lands). The collision between growing human populations, rural landowners, and associated culture and vast wildlife habitat needs requires consistent collaboration and stakeholder involvement. The definition of stakeholders is highly dependent on the project or situation in question but in general refers to anyone impacted by management actions (Cox 2013).

Involvement can also mean a variety of different things in this context; however, stakeholders are demanding greater influence in management decisions. "Decision space" refers to the ability of stakeholders to actually have concrete input and impacts on management plans and comes from a decision authority, which is usually a state or federal agency (Depoe, Delicath, and Elsenbeer 2004). Those with decision authority decide how much of this authority they can share with other stakeholders (the decision space). It is a difficult endeavor for both the entity with decision authority and the other stakeholders to agree on decision space.

Robust decision space means people and organizations involved in the process will have greater voice in the decision-making process, making them more likely to support the management actions. After all, they had a hand in crafting them. Certainly, stakeholder involvement has been mandated in federal policies such as the National Environmental Policy Act (NEPA). These mandates have uneven records for success, as they struggle with identifying stakeholders and providing decision authority to participants, or the situation is just too volatile to allow for stakeholder collaboration. However, the overarching benefit is that involvement of stakeholders brings more diversity to the process and highlights potential avenues of improvement. This requires identifying and engaging stakeholders such as agencies, nongovernmental organizations, landowners, and local citizens throughout the decision-making process. Much of the challenge is identifying relevant stakeholders and effectively sharing decision-making authority in a process that creates effective management plans. Without effective leadership, multistakeholder processes can struggle to make positive progress.

Proactive stakeholder involvement with effective leadership can mitigate, though rarely eliminate, conflict in habitat management. Conflict management is an important component of stakeholder involvement. However, disagreements are expected and can be beneficial to the process by allowing for a more comprehensive view of the issue from multiple perspectives (Mouffe 2000). The argumentative model can promote restructuring of extant power paradigms and allow new ideas to germinate (Peterson, Peterson, and Peterson 2005). Leadership and stakeholder involvement can be improved by hiring experienced facilitators to promote discourse between stakeholders, redirecting these initial tensions into productive, critical dialogue to create better management plans. There are a variety of approaches to identifying and involving stakeholders in management processes. We briefly focus on a few of the common methods: (1) public hearings, (2) public comment periods, (3) citizen advisory committees, and (4) community-based collaborations (Cox 2013).

Public Hearings

Public hearings are gatherings organized by entities with decision authority and designed to allow stakeholders to express their opinions on a proposed action. These are generally located in the areas where the action is expected to take place and involve key proposers of the action, such as state or federal agencies or representatives from private industry. For instance, US Forest Service representatives may hold a public hearing to discuss new timber extraction guidelines in a nearby national forest. Local community members, timber company personnel, and representatives of nongovernmental organizations and local municipalities may show up to present comments in response to the proposed new guidelines. Formats for speaking differ from meeting to meeting based on expected number of attendees and the level of controversy associated with the action.

Cox (2013) explained many of the problems concerning public hearings, including that (1) some stakeholders may not feel comfortable expressing opinions in public, (2) wait times to speak can be very long, (3) meeting times often conflict with work schedules, and (4) meetings can be very crowded and emotionally charged. Further critiques charge that public hearings are often little more than rituals that proposers must endure to ensure the proposed action. This is often referred to as decide, announce, defend. This paradigm provides little decision space to attendees and can often exacerbate tensions. Additionally, these tend to favor a top-down, technocratic approach where action proposers attempt to convince the audience of the benefits of the action (Peterson and Feldpausch-Parker 2013). Ultimately, the stakeholders are often left with little tangible evidence that their concerns or ideas actually impacted the process.

Public Comment Periods

Public comment periods are often closely associated with laws that mandate public participation, such as the NEPA. These can take multiple forms, including formal written comments or recorded audio at public meetings. The idea is that the public can inform decision makers about their concerns and preferences, and these comments will be taken into account during the decision making. For example, the US Fish and Wildlife Service (USFWS) solicits public comments for a set period of time through the *Federal Register* when it proposes to list a new species as endangered under the Endangered Species Act. The USFWS received over 600,000 comments when it proposed to list the polar bear (*Ursus maritimus*) as threatened (Inkley, Staudt, and Duda 2009). Although there are examples where this process has worked, these public comment periods rarely satisfy stakeholders' demands for inclusion and decision space (Peterson and Feldpausch-Parker 2013). In fact, both NEPA and the Aarhus Convention (which mandate public involvement in governmental decision making) lack any guidelines on how much weight public comment should actually be given. Ultimately, public comment periods limit stakeholder involvement in decision making when compared to other available methods.

Citizen Advisory Committees

The citizen advisory committee (CAC) is generally a governmentally appointed group designed to identify and receive input from stakeholders, for example, government-citizen committees that facilitate dialogue between the decision authorities and the other stakeholders. This effectively creates stakeholder involvement very early in the process, as citizens are a core part of these committees (Cox 2013). Additionally, these committees are selected to represent the actual stakeholders, including government, interest groups, and other citizens, with a final product being a set of recommendations for agency actions. A CAC might be associated with large conservation projects that impact multiple communities, such as implementation of conservation plans in watersheds or estuarine areas. A CAC made up of volunteer stakeholders can provide advice and feedback to federal and state regulators. In this way, the stakeholders have a tangible voice in the management plan.

Community-Based Collaborations

Cox (2013) described community-based collaborations as community-driven responses to natural resource conflicts. In an effort to reduce the loss of community resources and goodwill in acrimonious conflicts, agencies can create collaborative groups made up of stakeholders that address specific issues or conflicts in the community. These are often court appointed or supported by state and federal agencies and focus on short-term flash-fire conflicts that need cooperation. Although similar to CACs, community-based collaborations are often localized and focus on more specific problems or issues. Community-based collaborative models or strategies might also include collaborative learning (Daniels and Walker 2001), mediated modeling (Van den Belt 2004), and a participatory geographic information system (GIS; Elwood 2006).

Community-based collaborations allow for a high level of decision authority by stakeholders, thus creating recommendations to government agencies based on community input. An example might be a collaboration formed to manage the conversation surrounding recreational planning allowed or encouraged in a wilderness area. This topic may arise quickly and lead to heated debate among a variety of local stakeholders, such as hikers, campers, off-road vehicle aficionados, environmental groups, and others. A community-based collaboration initiative can bring these stakeholders together so they work (often with the help of a third-party facilitator) to determine a compromise recreational plan.

Conclusion

We have briefly described just a few methods for stakeholder involvement in decision making. It is impossible to go into great depth here, so we want to simply enlighten the reader to the importance of this topic. We highly recommend that readers interested in this subject consult Clarke and Peterson (2015); Cox (2013); and Depoe, Delicath, and Elsenbeer (2004). These methods have variable success and greatly depend on the ability of the decision authority to actually share decision-making power. This is further impacted by the quality of the process in how input is received and acted on and who is involved in the process. Even declared collaborative processes can become exclusionary as stakeholders are named and input is received. Arnstein's Ladder of Participation created a spectrum on which citizen participation can be judged (Arnstein 1969). Public hearings and public comments would rank relatively low on the scale (somewhere in tokenism), whereas CACs and community-based collaboration would rank much higher for effective participation.

What is not up for debate is the value of early and concerted efforts to involve stakeholders in these processes. For instance, the Texas A&M Institute of Renewable Natural Resources is involved in market-based wildlife habitat conservation in North Carolina, which needs private landowner participation. These are collaborations between the military, private landowners, and local communities to conserve red-cockaded woodpecker habitat on private lands to help the military in its dual mission of training personnel and conserving natural habitat. With effective incentivization and input from all stakeholders, not only do these stakeholders win but so does the endangered woodpecker. As the human population continues to grow, the value of these types of partnerships will only rise. Much of the wildlife habitat conservation in the future will need to take place on private lands because public lands are relatively limited in size and location. Additionally, these partnerships will be needed to form collectively beneficial management strategies such as water resource conservation and allocation in cities.

Summary

I. Prescribed Fire

A. Prescribed fire is used to manipulate habitat vegetation in land and wetland ecosystems.

B. Fire disturbance provides these benefits:

1. Management of woody vegetation and control of destructive pathogens
2. Improvement of watershed functions, water quality, and water quantity by controlling aquatic invasive plants and pathogens
3. Improvement of the palatability of forage and increase of their distribution

C. Prescribed fire is a more cost-effective option than other land management treatments.

II.   Mechanical Treatments

A. Mechanical treatment applications are land management tools used to manipulate vegetation and modify soil conditions.

1. Mechanical treatments can be used to reduce hazardous fuels and water runoff.
2. Mechanical treatments can be used to break up compacted soils.

B. Hand or power mechanical treatments vary in technique, cost, and application benefits and are largely determined by plant community, species density, terrain, and climatic conditions.

C. Achieving specific goals and results often requires mechanical treatments to be administered several times for proper vegetative control of the target.

III.   Grazing

A. Grazing is a land management tool that shifts the feeding intensity of specific areas/pastures by rotating livestock to increase production of wildlife and livestock.

1. Grazing is used to clear land and control invasive vegetation.
2. Grazing is a method to improve specific animal habitat.

B. Integrated grazing management systems require land managers to evaluate and assess wildlife and livestock grazing interactions and minimize overgrazing and decrease herbivory competition.

IV.   Harvest Management

A. Regulated harvesting of wild animals allows managers some control over populations and their impacts on habitat.

1. Regulated harvests can be adjusted to meet emergent population or habitat issues such as natural disasters, demographic or population changes, or land use alterations.
2. Potential benefits of such control include improved wildlife health and persistence (appropriate demographics, increased juvenile survival, decreased disease), habitat quality (preferred vegetation structures), and human-nature linkages.

B. Every proposed harvest must begin with clear and well-reasoned objectives that take both ecology and social constructs into account.

1. Accurate population and harvest data are critical to creating appropriate harvest strategies and must be consistent and available throughout the process.

2. Managers must be able to access immediate data, evaluate current conditions, and then alter management methods accordingly.

V.   Herbicide Applications

A. Herbicide application is a land management tool used to manipulate habitat vegetation in land and wetland ecosystems.

B. Some benefits of herbicide treatments include control of unwanted woody vegetation and destructive pathogens, improvement of forage quality and palatability, promotion of tree growth, and improvement of land accessibility and visibility.

VI.   Supplemental and Replacement Water

A. Management plans require careful consideration, monitoring, and maintenance; should identify wildlife water needs during times of stress such as drought; and provide water at appropriate times and in sufficient amounts.

B. When applied appropriately in a management framework with well-understood goals, supplemental and replacement water can improve animal production and health.

VII. Urban and Suburban Processes

A. Urban and suburban areas are dominated by human impacts but can still support wildlife.

B. Urban and suburban areas also require wildlife habitat management.

C. The persistence of urban wildlife habitat requires concerted and directed management through, for example, urban forestry and urban agriculture.

VIII. Stakeholder Involvement

A. Stakeholder involvement in decision making is critical to successful management.

B. In general, the term "stakeholder" refers to anyone impacted by the management actions in question.

C. Involvement of stakeholders can provide decision space, which means that people and organizations involved in the process will have greater voice in the decision-making process.

D. There are a variety of approaches to identifying and involving stakeholders in management processes, including but not limited to the following:

1. Public hearings
2. Public comment periods

3. Advisory committees
4. Community-based collaborations

E. Each method has positive and negative aspects; however, public hearings and comment periods are generally seen as less effective than the other methods.

Literature Cited

Agriculture Victoria. 2016. Top 10 spraying tips. http://agriculture.vic.gov.au/agriculture/farm-management/chemical-use/agricultural-chemical-use/spraying-spray-drift-and-off-target-damage/top-10-spraying-tips.

Alvey, A. A. 2006. Promoting and preserving biodiversity in the urban forest. *Urban Forestry and Urban Greening* 5:195–201.

Anderson, J. R., E. E. Hardy, J. T. Roach, and R. E. Witmer. 1976. *A land use and land cover classification system for use with remote sensor data*. Washington, DC: US Geological Survey.

Arnstein, S. R. 1969. A ladder of citizen participation. *Journal of the American Institute of Planners* 35:216–24.

Boal, C. W., and R. W. Mannan. 1999. Comparative breeding ecology of Cooper's hawks in urban and exurban areas of southeastern Arizona. *Journal of Wildlife Management* 63:77–84.

Bovey, R. W. 1998. *A fifty-year history of the weed and brush program in Texas and suggested future directions*. Bulletin B-1729. College Station: Texas Agricultural Experiment Station.

Bradley, C. A., and S. Altizer. 2007. Urbanization and the ecology of wildlife diseases. *Trends in Ecology and Evolution* 22:95–102.

Brennan, L. A., and W. P. Kuvlesky. 2005. North American grassland birds: An unfolding conservation crisis? *Journal of Wildlife Management* 69:1–13.

Brenneisen, S. 2006. Space for urban wildlife: Designing green roofs as habitats in Switzerland. *Urban Habitats* 4:27–36.

Bussan, A. J., and W. E. Dyer. 1999. Herbicides and rangelands. In *Biology and management of noxious rangeland weeds*, edited by R. L. Sheley and J. K. Petroff, 116–32. Corvallis: Oregon State University Press.

Carpenter, L. H. 2000. Harvest management goals. In *Ecology and management of large mammals in North America*, edited by S. Demarais and P. R. Krausman, 192–213. Upper Saddle River, NJ: Prentice-Hall.

Caula, S. G., T. Hvenegaard, and P. Marty. 2009. The influence of bird information, attitudes, and demographics on public preferences toward urban green spaces: The case of Montpellier, France. *Urban Forestry and Urban Greening* 8:117–28.

Clark Conservation District. 2013. Pasture systems and grazing methods. http://static1.squarespace.com/static/52af6c52e4b035961e75998b/t/52b09dade4b0324077fb5491/1387306413094/Pasture+Systems+and+Grazing+Methods2.pdf.

Clark, J. R., N. P. Mathey, G. Cross, and V. Wake. 1997. A model of urban forest sustainability. *Journal of Arboriculture* 23:17–30.

Clarke, T., and T. R. Peterson. 2015. *Environmental conflict management*. Washington, DC: Sage.

Clucas, B., J. M. Marzluff, S. Kübler, and P. Meffert. 2011. New directions in urban avian ecology: Reciprocal connections between birds and humans in cities. In *Perspectives of urban ecology*, edited by W. Endlicher, 167–95. Heidelberg, Germany: Springer-Verlag.

Conner, J. R., G. W. Williams, and R. A. Dietrich. 2003. Cattle and the environment: What's the beef? *Proceedings of Western Economics Forum* 2:3–7.

Cook, C. R., and L. E. Harris. 1968. *Nutritional value of seasonal ranges*. Bulletin 472. Logan: Utah Agricultural Experiment Station.

Corbett, E. S., J. A. Lynch, and W. E. Sopper. 1978. Timber harvesting practices and water quality in the eastern United States. *Journal of Forestry* 76:484–88.

Cox, R. 2013. *Environmental communication and the public sphere*. 3rd ed. Thousand Oaks, CA: Sage.

Crocker-Bedford, D. C. 1990. Goshawk reproduction and forest management. *Wildlife Society Bulletin* 18:262–69.

Daniels, S. E., and G. B. Walker. 2001. *Working through environmental conflict: The collaborative learning approach*. Westport, CT: Praeger.

Dearborn, D. C., and S. Kark. 2010. Motivations for conserving urban biodiversity. *Conservation Biology* 2:432–40.

Decker, D. J., and N. A. Connelly. 1989. Motivations for deer hunting: Implications for antlerless deer harvest as a management tool. *Wildlife Society Bulletin* 17:455–63.

De Fries, R. S., J. A. Foley, and G. P. Asner. 2004. Land-use choices: Balancing human needs and ecosystem function. *Frontiers of Ecology and the Environment* 2:249–57.

Depoe, S. P., J. W. Delicath, and M. A. Elsenbeer. 2004. *Communication and public participation in environmental decision making*. Albany: State University of New York Press.

Elwood, S. 2006. Critical issues in participatory GIS: Deconstructions, reconstructions, and new research directions. *Transactions in GIS* 10:693–708.

Federal Aviation Administration. 2008. *Pilot's handbook of aeronautical knowledge*. FAA-H-8083-25A. Washington, DC: US Department of Transportation.

Fernandez-Giménez, M. E, and D. M. Swift. 2003. Strategies for sustainable grazing management in the developing world. In *Rangelands in the new millennium, VIIth International Rangelands Congress*, 26 July–1 August 2003, Durban, South Africa, edited by N. Allsopp and N. Walker, 821–31. Durban, South Africa: The Congress.

Fischer, J., B. Brosi, G. C. Daily, P. R. Ehrlich, R. Goldman, J. Goldstein, D. B. Lindenmayer, A. D. Manning, H. A. Mooney, L. Pejchar, J. Ranganathan, and H. Tallis. 2008. Should agricultural policies encourage land sparing or wildlife-friendly farming? *Frontiers of Ecology and the Environment* 6:380–85.

Frank, D. A., S. J. McNaughton, and B. F. Tracy. 1998. The ecology of the earth's grazing ecosystems. *BioScience* 48:513–21.

Franklin, J. F., T. A. Spies, R. Van Pelt, A. B. Carey, D. A. Thornburgh, D. R. Berg, D. B. Lindenmayer, M. E. Harmon, W. S. Keeton, D. C. Shaw, K. Bible, and J. Chen. 2002. Disturbances and structural development of natural forest ecosystems with silvicultural implications, using Douglas-fir forests as an example. *Forest Ecology and Management* 155:399–423.

Freilich, J. E., J. M. Emlen, J. J. Duda, D. C. Freeman, and P. J. Cafaro. 2003. Ecological effects of ranching: A six-point critique. *BioScience* 52:759–65.

Fulbright, T. E. 1996. A theoretical basis for planning woody plant control to maintain species diversity. *Journal of Range Management* 49:554–59.

Gibbs, J. P. 1993. Importance of small wetlands for the persistence of local populations of wetland-associated animals. *Wetlands* 13:25–31.

———. 2000. Wetland loss and biodiversity conservation. *Conservation Biology* 14:314–17.

Green, R. E., S. J. Cornell, J. P. W. Scharlemann, and A. Balmford. 2005. Farming and the fate of nature. *Science* 307:550–55.

Hankins, D. L. 2005. Pyrogeography: Spatial and temporal relationships of fire, nature, and culture. PhD diss., University of California, Davis.

Harrison, C., and G. Davies. 2002. Conserving biodiversity that matters: Practitioner's perspectives on brownfield development and urban nature conservation in London. *Journal of Environmental Management* 65:95–108.

Haufler, J. B., and A. C. Ganguli. 2007. *Benefits of farm bill grassland conservation practices to wildlife*. The Wildlife Society Technical Review 07-1. http://www.nrcs.usda.gov/Internet/FSE_DOCUMENTS/nrcs143_013148.pdf.

Heitschmidt, R. K., L. T. Vermeire, and E. E. Grings. 2004. Is rangeland agriculture sustainable? *Journal of Animal Science* 82:138–46.

Helms, J. 1998. *The dictionary of forestry*. Bethesda, MD: Society of American Foresters.

Hibbert, A. R. 1983. Water yield improvement potential by vegetation management on western rangelands. *Water Resources Bulletin* 19:375–81.

Holechek, J. L. 1980. Concepts concerning forage allocation to livestock and big game. *Rangelands* 2:158–59.

Horsley, S. B., S. L. Stout, and D. S. deCalesta. 2003. White-tailed deer impact on the vegetation dynamics of a northern hardwood forest. *Ecological Applications* 13:98–118.

Inkley, D. B., A. C. Staudt, and M. D. Duda. 2009. Imagining the future: Humans, wildlife, and global climate change. In *Wildlife and society: The science of human dimensions*, edited by M. J. Manfredo, J. J. Vaske, P. J. Brown, D. J. Decker, and E. A. Duke, 57–72. Washington, DC: Island Press.

Kellermann, J. L., M. D. Johnson, A. M. Stercho, and S. C. Hackett. 2008. Ecological and economic services provided by birds on Jamaican Blue Mountain coffee farms. *Conservation Biology* 22:1177–85.

Kilgo, J. C., R. F. Labisky, and D. E. Fritzen. 1998. Influences of hunting on the behavior of white-tailed deer: Implications for conservation of the Florida panther. *Conservation Biology* 12:1359–64.

Krueger, M., and J. Dillard. 2013. *Developing a deer management plan*. Texas Parks and Wildlife Department. http://www.tpwd.state.tx.us/publications/pwdpubs/media/pwd_rp_w7000_1132.pdf.

Leopold, A. 1933. *Game management*. Madison: University of Wisconsin Press.

Lundkvist, A. 1997. Predicting optimal application time for herbicides from estimated growth rate of weeds. *Agricultural Systems* 54:223–42.

Mander, Ü., M. Mikk, and M. Külvik. 1999. Ecological and low intensity agriculture as contributors to landscape and biological diversity. *Landscape and Urban Planning* 46:169–77.

McClintock, N. 2010. Why farm the city? Theorizing urban agriculture through a lens of metabolic rift. *Cambridge Journal of Regions, Economy and Society* 3:191–207.

McCullough, D. R. 1996. Spatially structured populations and harvest theory. *Journal of Wildlife Management* 60:1–9.

McKinney, M. L. 2002. Urbanization, biodiversity, and conservation. *BioScience* 52:883–90.

McPherson, E. G. 2006. Urban forestry in North America. *Renewable Resources Journal* 24:8–12.

Miller, J. R. 2007. Habitat and landscape design: Concepts, constraints and opportunities. In *Landscapes for conservation: Moving from perspectives to principles*, edited by D. Lindenmayer and R. J. Hobbs, 81–95. Malden, MA: Blackwell Publishing.

Mitsch, W. J., and J. G. Gosselink. 2000. *Wetlands*. 3rd ed. New York: John Wiley and Sons.

Miyamoto, D., R. Olson, and G. Schuman. 2004. Long-term effects of mechanical renovation of a mixed-grass prairie: I. Plant production. *Arid Land Research and Management* 18:93–101.

Monsanto. 2009. *Herbicide application handbook: A guide to proper handling and application of Monsanto herbicides*. St. Louis, MO: Monsanto.

Morrison, M. L., R. C. Heald, and D. L. Dahlsten. 1990. Can incense-cedar be managed for birds? *Western Journal of Applied Forestry* 5:28–30.

Mouffe, C. 2000. *The democratic paradox*. London: Verso.

Natural Resources Conservation Service (NRCS). 2003. *National range and pasture handbook*. Revision 1. Washington, DC: US Department of Agriculture, Grazing Lands Technology Institute.

Nelson, A. C. 1992. Preserving prime farmland in the face of urbanization: Lessons from Oregon. *Journal of the American Planning Association* 58:467–88.

North Dakota State University Cooperative Extension. 2013. Herbicide use during cold weather. https://www.ag.ndsu.edu/winterstorm/winter-storm -information-farm-and-ranch-information/farm-and-ranch-crops-general/ herbicide-use-during-cold-weather.

Nowak, D. J. 2006. Institutionalizing urban forestry as "biotechnology" to improve environmental quality. *Urban Forestry and Urban Greening* 5:93–100.

Parker, I. D., and A. M. Feldpausch-Parker. 2013. Yellowstone grizzly delisting rhetoric: An analysis of the online debate. *Wildlife Society Bulletin* 37:248–55.

Patterson, M. E., J. M. Montag, and D. R. Williams. 2003. The urbanization of wildlife management: Social science, conflict, and decision-making. *Urban Forestry and Urban Greening* 1:171–83.

Peterson, M. N. 2004. An approach for demonstrating the social legitimacy of hunting. *Wildlife Society Bulletin* 32:310–21.

Peterson, M. N., H. P. Hansen, M. J. Peterson, and T. R. Peterson. 2010. How hunting strengthens social awareness of coupled human-natural systems. *Wildlife Biology Practice* 6:127–43.

Peterson, M. N., M. J. Peterson, and T. R. Peterson. 2005. Conservation and the myth of consensus. *Conservation Biology* 19:762–67.

Peterson, T. R., and A. M. Feldpausch-Parker. 2013. Environmental conflict communication. In *The SAGE Handbook on Conflict Communication: Integrating theory, research, and practice*, 2nd ed., edited by J. G. Oetzel and S. Ting-Toomey, 513–35. Thousand Oaks, CA: Sage.

Pike, D. A., J. K. Webb, and R. Shine. 2011. Chainsawing for conservation: Ecologically informed tree removal for habitat management. *Ecological Management and Restoration* 12:110–18.

Provencher, L., N. M. Gobris, L. A. Brennan, D. R. Gordon, and J. L. Hardesty. 2002. Breeding bird response to midstory hardwood reduction in Florida sandhill longleaf pine forests. *Journal of Wildlife Management* 66:641–61.

Quon, S. 1999. *Planning for urban agriculture: A review of tools and strategies for urban planners*. Cities Feeding People Series, Report 28. Ottawa, Ontario: International Development Research Centre.

Rickel, B. 2005. Large native ungulates. In *Assessment of grassland ecosystem conditions in the southwestern United States: Wildlife and fish*, vol. 2, edited by D. M. Finch, 13–34. Albuquerque, NM: US Forest Service, Rocky Mountain Research Station.

Robinson, J. G., and E. L. Bennett. 2004. Having your wildlife and eating it too: An analysis of hunting sustainability across tropical ecosystems. *Animal Conservation* 7:397–408.

Rudd, H., J. Vala, and V. Schaefer. 2002. Importance of backyard habitat in a comprehensive conservation strategy: A connectivity analysis of urban green spaces. *Restoration Ecology* 10:368–75.

Sargent, M. S, and K. S. Carter, eds. 1999. *Managing Michigan wildlife: A landowner's guide*. East Lansing: Michigan United Conservation Clubs.

Sauer, C. D. 1950. Grassland climax, fire, and man. *Journal of Range Management* 3:16–21.

Scifres, C. J., and H. T. Hamilton. 1993. *Prescribed burning for brushland management: The South Texas example*. College Station: Texas A&M University Press.

Shore, R. F., W. R. Meek, T. H. Sparks, R. F. Pywell, and M. Nowakowski. 2005. Will environmental stewardship enhance small mammal abundance on intensively managed farmland? *Mammal Review* 35:277–84.

Smith, B. W., P. D. Miles, J. S. Vissage, and S. A. Pugh. 2002. *Forest resources of the United States, 2002*. St. Paul, MN: US Forest Service, North Central Research Station.

Southerton, N. W. 1998. Land use changes and the decline of farmland wildlife: An appraisal of the set-aside approach. *Biological Conservation* 83:259–68.

Strickland, B., S. Edwards, and R. Hamrick. 2016. *Prescribed burning in southern pine forests: Fire ecology, techniques, and uses for wildlife management*. Mississippi State University Extension Service, Publication 2283 (POD-03-16). Starkville: Mississippi State University Extension Service.

Texas Parks and Wildlife Department. 2007. *Wildlife management activities and practices: Comprehensive wildlife management planning guidelines for the post oak savannah and blackland prairie ecological regions*. Austin: Texas Parks and Wildlife Department.

Theobold, D. M., J. R. Miller, and N. T. Hobbs. 1997. Estimating the cumulative effects of development on wildlife habitat. *Landscape and Urban Planning* 39:25–36.

Treadaway, M., V. W. Howard Jr., C. D. Allison, and J. C. Boren. 1997. Elk vs. livestock: Forage utilization study in portions of the Gila National Forest. *Proceedings of the Great Plains Wildlife Damage Control Workshop.* Internet Center for Wildlife Damage Management Paper 379. http://digitalcommons.unl.edu/cgi/viewcontent.cgi?article=1378&context=gpwdcwp.

Turrentine, J. 1992. *Appendix O: Wildlife watering facilities.* USDA/NRCS Tech Note Biology TX-19. https://tpwd.texas.gov/landwater/land/private/agricultural_land/stx2010/Appendix%200%20Wildlife%20Watering%20Facilities.pdf.

Van den Belt, M. 2004. *Mediated modeling: A system dynamics approach to environmental consensus building.* Washington, DC: Island Press.

Volesky, J. D., J. L. Stubbendieck, and R. B. Mitchell. 2007. *Conducting a prescribed burn and prescribed burning checklist.* University of Nebraska Cooperative Extension EC121. Lincoln: University of Nebraska Cooperative Extension.

Walburger, K., T. Delcurto, M. Vavra, L. Bryant, and J. G. Kie. 2000. Influence of a grazing system and aspect, north vs. south, on the nutritional quality of forages, and performance and distribution of cattle grazing forested rangelands. In *Proceedings of the Western Section, American Society of Animal Science,* June 20–23, 2000, Davis, CA.

Wegner, J. F., and G. Merriam. 1979. Movements by birds and small mammals between a wood and adjoining farmland habitats. *Journal of Applied Ecology* 16:349–57.

5 Wildlife Habitat Planning

> I have read many definitions of what is a conservationist, and written not a
> few myself, but I suspect the best one is written not with a pen, but with an
> axe. . . . A conservationist is one who is humbly aware that with each stroke
> he is writing his signature on the face of the land. Signatures of course
> differ, whether with axe or pen, and this is as it should be.
> —Aldo Leopold, *Sand County Almanac*

The quotation by Aldo Leopold illustrates the importance of the careful
execution of habitat management to "do the right thing" with regard
to enhancing and protecting natural resources. Thus far, we have reviewed basic principles in wildlife-habitat relationships, approaches to
measuring and evaluating wildlife habitat, and common habitat modification practices. The integration of theory and practice is typically
accomplished through conservation and management planning. This
chapter describes both the process of conservation planning and its
associated management plans. Conservation planning is a creative process that requires experience, analysis, intuition, and even inspiration.
Management plans can add structure to and are an integral part of the
overall conservation planning process (Lopez et al. 2006). Conservation
planning is a progression of steps through which we determine *what
we have* (animals, plants, physical resources, etc.), *what we want* (e.g.,
more or fewer white-tailed deer), and *how to get there* (e.g., increase or
decrease cover or forage requirements). Although there is no single
method or cookie-cutter approach to developing a conservation plan,
there are general guidelines that can aid in successfully engaging in the
process and increase awareness of a manager's "signature on the face of
the land." The following sections introduce basic concepts in the development and careful execution of management plans.

What Is Planning?

A good working definition of planning is "the deliberate social or organizational activity of developing an optimal strategy for solving problems and achieving a desired set of objectives" (Yoe and Orth 1996, 10).

Although this definition oversimplifies an often complex process, it does emphasize a few important elements of planning:

1. *Future control.* Planning is focused on the control of future consequences through present actions. Both planning and action are necessary to ensure increased control of the targeted lands in the future, including minimizing or eliminating any uncertainty that invariably will be part of the planning process.

2. *Problem solving.* Planning is a problem-solving approach that addresses a particular natural resource issue. Generally, wildlife management problems could include increasing population numbers for endangered species, decreasing impacts from nuisance or invasive species, or maintaining a sustained yield for a game species. Thus, conservation planning often is driven by the need to address such a problem or natural resource issue.

3. *Team effort.* Planning is often done by individuals in a team environment that considers the opinions and desires of a wide-ranging set of stakeholders, such as landowners or state and local governments. Who is involved depends on the spatial extent of the property and the landowner; for example, a plan for a small private property may be done by an individual. Planning should not be done without considering various perspectives and approaches. Increasing the level of input in conservation planning can ensure the careful review of all viable options.

4. *No single approach.* Planning is uniquely tailored to a specific situation and set of objectives and assumptions. Unfortunately, there is no single approach in conservation planning. General rules and procedures can be applied, however, and the manager should be willing to adapt and modify the plan to the specific area to improve its overall effectiveness.

5. *Adaptive framework.* Planning is adaptive and involves "feedback" loops. Monitoring, evaluation, and adjustment are critical components of good management plans, which are part of a process. This process typically has a predetermined goal that is reevaluated at the end of the target time line and restarts the management process.

6. *Intention to implement.* Planning is done with the intention of implementing the strategies outlined in the plan, not just for planning's sake. Management plans require a substantial amount of time, effort, and funding. Land managers should develop management plans with the intention of following them, not simply put them on a shelf.

Goals and Objectives

Our working definition of planning includes a strategy for achieving a desired set of objectives. Management plans are very much objective and goal oriented. The terms "goals" and "objectives" are sometimes used interchangeably; however, there are subtle differences when used in relation to conservation planning. According to *Webster's*, a *goal* is "the end or final purpose." Conversely, an *objective* is "something aimed at or striven for." From a conservation planning perspective, both definitions convey the same basic intent: "do the right thing" for the resource. But the subtle differences in their definitions suggest that we set goals first and then establish objectives to attain the goals. A goal relates to "doing the right thing" *broadly*, whereas an objective relates *specifically*. The following example differentiates the two from the perspective of two land managers who share the common goal of improving quail habitat but differ in how they individually will strive to reach this goal.

Land Manager 1

- Goal: Improve quail habitat
- Objective 1: Apply prescribed fire
- Objective 2: Disk annually
- Objective 3: Maintain brush piles

Land Manager 2

- Goal: Improve quail habitat
- Objective 1: Control predators
- Objective 2: Plant food plots
- Objective 3: Increase native grasses

Each land manager has the same goal, yet the objectives are very different. In both cases, the managers' individual objectives allow them to reach their common goal. As Leopold so eloquently said, "Signatures of course differ, whether with axe or pen, and this is as it should be."

Types of Planning

There are many different types of plans. *Land use* plans attempt to maintain the most suitable types of land uses, such as developed versus natural areas, and specific activities relevant to the management of those lands. *Transportation* plans may address public transportation strategies to support the projected growth of a region or county. *Historic preservation* plans might be used in a community to protect a historic area like a Civil War battlefield or the remnants of an old mining community. These are a few examples of the many types of plans managers are likely to come across in their experiences with conservation planning. A common challenge is the lack of coordination and integration among these planning processes within the same geographic area.

Conflicting approaches and recommendations between transportation and conservation plans, for example, require a concerted effort for agencies and other resource managers to work collaboratively. This challenge arises in large part because various agencies and organizations are responsible in the management of natural resources (table 5.1). Organizations or agencies responsible for conservation planning can range from federal and state agencies to nonprofit organizations, land trusts, and other local/regional partnerships. That so many stakeholders are involved in the management of natural resources can be challenging and even daunting at times and can make it difficult to determine responsibilities. This is exacerbated by the common phenomenon of mission creep (projects expanding beyond or changing from original parameters).

Planning can also vary in spatial scale or a particular focus. For example, conservation plans can range from multistate ecosystem-level plans (e.g., America's Longleaf Restoration Initiative) to local habitat conservation plans that focus on a single endangered species. All levels of planning are important, but each individual plan is likely to be more effective when it connects with other planning efforts. Including information from a local habitat conservation plan in an ecoregional plan maintains consistency and increases coordination, as was the case in the development of The Nature Conservancy's (TNC) ecoregional assessments. TNC realized expenditures of its funding would be more effective if it developed a long-range plan for land acquisition and restoration based on the best available science and carried it out at a larger, regional scale. As a result, TNC organized planning by ecoregions, which in some cases span multiple states and counties, and identified and prioritized target conservation land based on the current ecology and threat levels within that region (table 5.2). This is an example of how conservation planning can be integrated and coordinated from the local to landscape level.

Table 5.1. Agencies and organizations responsible for conservation planning

Federal agencies	State agencies	Municipal government	Nonprofit organizations	Local land trusts
Army Corps of Engineers	**Maine**	Conservation commissions	American Farmland Trust	Five Rivers Land Trust
Bureau of Land Management	Maine Bureau of Parks and Lands	Open space committees	Audubon Society of North Carolina	Jericho Underhill Land Trust
Department of Defense	Maine Department of Inland Fisheries and Wildlife	Town forest committees	Audubon Vermont	Lake Champlain Land Trust
National Parks Service	Maine Forest Service		Forest Society of Maine	Lakes Region Conservation Trust
Natural Resources Conservation Service			Maine Audubon Society	Lower Kennebec Regional Land Trust
US Fish and Wildlife Service	**Florida**		Society for the Protection of New Hampshire Forests	Mahoosuc Land Trust
US Forest Service	Florida Department of Environmental Protection		Texas Land Trust	Maine Wilderness Watershed Trust
	Florida Fish and Wildlife Conservation Commission		The Conservation Fund	Middlebury Area Land Trust
	Florida Park Service		The Nature Conservancy	Mondadnock Conservancy
			Trust for Public Lands	Piscataquog River Watershed Association
	Texas			Sebasticook River Watershed Association
	Texas Forest Service			
	Texas Parks and Wildlife Department			

Table 5.2. Examples of conservation planning efforts by scale or focus

Scale or focus	Example	Description
Landscape	America's Longleaf Restoration Initiative	America's Longleaf Restoration Initiative is a voluntary collaborative effort by more than 20 organizations and agencies that seeks to define, catalyze, and support coordinated longleaf pine conservation efforts across a nine-state region.
State	State wildlife action plans	State wildlife action plans take a proactive approach to habitat conservation and species preservation. These plans outline a strategy for protecting priority habitats and species that are at risk but not yet on the endangered species list in conjunction with state wildlife agencies from all 50 states, all US territories, and the District of Columbia.
Regional	The Nature Conservancy's ecoregional assessments	The Nature Conservancy's long-range plan for land acquisition and restoration is based on natural ecoregions. Each ecoregional assessment is identified and prioritizes conservation lands based on ecology and threat levels.

Major Legislation and Policy Directing Wildlife Planning

Another question to consider is "why plan?" In addition to previously mentioned benefits of conservation planning, three basic frameworks commonly drive conservation planning: (1) agency directives or authorities (laws or policy that drives an organization's mission), (2) environmental compliance (regulatory laws), and (3) incentive programs such as Farm Bill programs. Most conservation *funding* to support implementation of management activities outlined in conservation plans typically requires a management plan of some sort in order to receive and use that funding. A summary of the various federal and state conservation funding programs illustrates the value and reason for developing management plans (table 5.3). Familiarity with these various programs, which are described in the following sections, is important to aid in obtaining financial support for conservation or, if working with private landowners, encourages certain behaviors that can benefit conservation and wildlife populations.

Table 5.3. Summary of federal and state conservation funding programs

Program	Agency	Description
Readiness and Environmental Protection (REPI)	US Department of Defense (DoD)	Cost-sharing program for the acquisition of easements from willing sellers as a way to preserve high-value habitat and limit incompatible development around military ranges and installations.
Land and Water Conservation Fund Stateside Program (LWCF)	US Department of the Interior–National Park Service (DOI-NPS)	State matching grant program (at least 50%) where states request funds from the National Park Service for specific projects.
Land and Water Conservation Fund Federal Land Acquisition (LWCF)	US Department of the Interior–National Park Service (DOI-NPS)	Program to acquire new federal recreation lands in cooperation with the National Park Service.
Urban Parks and Recreation Recovery (UPARR)	US Department of the Interior–National Park Service (DOI-NPS)	Grants to urban communities for rehabilitation of facilities, parks planning, and innovative programs. FY2002 is the last year that grants were awarded.
Transportation Enhancements (TE)	US Department of Transportation (DOT)	The TE program represents 10% of total Surface Transportation Program (STP) funds. STP is a formula for apportionment (maximum 80%) to the states through the Federal-Aid Highway Program. TE projects include construction, but not maintenance, of various modes of transportation, including the rail trail program, which is funded by an excise tax from the Highway Trust Fund. Projects involve conversion of abandoned railroad corridors into multiuse trails available for recreation.
Recreational Trails Program (RTP)	US Department of Transportation (DOT)	Apportionments (maximum 80%) to the states to benefit outdoor recreation, including hiking, biking, in-line skating, equestrian use, etc.
Forest Legacy Program (FLP)	US Forest Service (FS)	Competitive grant program and direct payments to support land acquisition (fee purchase and easement) to protect important scenic, cultural, fish, wildlife, and recreation resources as well as riparian areas.
Forest Stewardship	US Forest Service (FS)	Provides technical and educational assistance to nonindustrial private forest owners to develop forest management plans.
Forestland Enhancement Program	US Forest Service (FS)	Complements Forest Legacy with additional technical, educational, and cost-share assistance to nonindustrial private forest owners to implement management plans. Only $40 million of original $100 million budgeted was actually distributed, all by FY2006. The program was not renewed in the 2008 Farm Bill.
Urban and Community Forestry	US Forest Service (FS)	Forest-related technical, financial, research, and educational services to local governments, nonprofits, community groups, and educational institutions.
Community Forest and Open Space Conservation Program (CFOSCP)	US Forest Service (FS)	Matching grants (50/50) for local governments, tribes, and nonprofit organizations for full fee purchase of forestlands; differs from Forest Legacy in its community focus and the requirement of fee purchase plus public access. The program is part of the 2008 Farm Bill; no appropriations yet.

Program	Agency	Description
Cooperative Endangered Species Conservation Fund	US Fish and Wildlife Service (FWS)	Grants to private landowners and groups to implement conservation projects for listed species and at-risk species. Funded activities include developing Habitat Conservation Plans, land acquisition, habitat restoration, research, and wildlife management.
Migratory Bird Conservation Fund	US Fish and Wildlife Service (FWS)	Land and water acquisition or rental as recommended by the secretary of Interior for the protection of migratory bird species. Funding comes from Duck Stamp revenues, import duties on arms and ammunition, and refuge admission fees.
National Coastal Wetlands Conservation Grants	US Fish and Wildlife Service (FWS)	Matching grants to states for acquisition, restoration, and enhancement of coastal wetlands. Funding comes from the Sport Fish Restoration and Boating Trust Fund, which is supported by excise taxes on fishing equipment, motorboat and small-engine fuels, and import duties.
North American Wetlands Conservation Grants (NAWCA)	US Fish and Wildlife Service (FWS)	Matching grants to organizations and individuals to implement wetlands conservation projects in the United States, Canada, and Mexico. Funding comes from congressional appropriations as well as fines and penalties collected under the Migratory Bird Treaty Act of 1918, the Sport Fish Restoration and Boating Trust Fund, and interest accrued on the Wildlife Restoration Trust Fund.
Sport Fish Restoration Program (SFR)	US Fish and Wildlife Service (FWS)	Apportionments to the states for fishery projects, boating access, and aquatic education. Funded by the Sport Fish Restoration and Boating Trust Fund, which is supported by excise taxes on fishing equipment, motorboat and small-engine fuels, and import duties. The annual allocation to the Sport Fish Restoration Program from Dingell-Johnson Act or the Wallop-Breau Act is equal to 57% of the trust fund's receipts (after annual deductions).
Wildlife Restoration Program (WRP)	US Fish and Wildlife Service (FWS)	Apportionments to the states to restore, conserve, and manage wild birds and mammals and their habitat. This program is funded by the Wildlife Restoration Trust Fund, which is supported by excise taxes, authorized by the Pittman-Robertson Act, on hunting equipment.
State Wildlife Grant Program	US Fish and Wildlife Service (FWS)	Grants to plan and implement programs that benefit wildlife and habitats, including species not hunted or fished. Funding comes through appropriations from the Land and Water Conservation Fund.
Landowner Incentive Program (LIP)	US Fish and Wildlife Service (FWS)	State grants to protect and restore habitats on private lands to benefit at-risk species (including federally listed, proposed, or candidate species). The program was discontinued after FY2007.
National Fish Habitat Action Plan (NFHAP)	US Fish and Wildlife Service (FWS) and others	Currently includes restoration activities only, but on congressional approval of National Fish Habitat Conservation Act, it will include land acquisition.
National Estuarine Research Reserves Program (NERRS)	National Oceanic and Atmospheric Administration (NOAA)	Grants to coastal states to acquire lands and waters necessary to ensure long-term management of an area as a national estuarine reserve and for operations, construction, and education programs.

Directives / Agency Guidance

Federal and state agencies all work under some sort of authority or set of laws and/or policies in support of the organizations' mission. Non-governmental organizations may not necessarily work under a given authority but may either be driven or impacted by some federal or state law, such as the Endangered Species Act, or directly benefit from authorities that provide financial or technical assistance. For example, the Federal Aid in Wildlife Restoration Act (more commonly known as the Pittman-Robertson Act), was approved by Congress in 1937 to "provide funding for the selection, restoration, rehabilitation and improvement of wildlife habitat, wildlife management research, and the distribution of information produced by the projects" (16 U.S.C. 669–669i; 50 Stat. 917). The Pittman-Robertson Act was amended in 1970 to include funding for hunter training programs and the development, operation, and maintenance of public target ranges.

Funds are determined from an 11% federal excise tax on sporting arms, ammunition, and archery equipment and a 10% tax on handguns. These funds are collected from the manufacturers and apportioned annually to the states and appropriate US territories by the Department of the Interior. Funds are distributed to states using a predetermined formula, which considers the total area of the state and the number of licensed hunters in the state. Appropriate state agencies are the only entities eligible to receive Federal Aid grant funds. This is a cost-reimbursement program, which means the state covers the full amount of an approved project and then applies for reimbursement for up to 75% of the project expenses. The state must provide at least 25% of the project costs from a nonfederal source. To receive Pittman-Robertson funds, a state must submit a comprehensive fish and wildlife resource management plan (e.g., State Wildlife Action plan). The plan must be laid out for at least five years and must be based on long-range projections regarding the desires and needs of the public.

Similarly, the Federal Aid in Sport Fish Restoration Act (also known as the Dingell-Johnson Act) was passed in 1950 and modeled after the Pittman-Robertson Act. The purpose of the Dingell-Johnson Act was to create a parallel program for management, conservation, and restoration of fishery resources. The Sport Fish Restoration program is funded by revenues collected from the manufacturers of fishing rods, reels, creels, lures, flies, and artificial baits. The program also is a cost-reimbursement program (75% federal, 25% nonfederal state match), and as necessary to receive Pittman-Robertson funds, each state must submit a five-year comprehensive fish and wildlife resource management plan to receive funding. In the expenditure of both Pittman-Robertson and Dingell-Johnson funding, management plans are critical components in state conservation agencies' obtaining and supporting their wildlife programs through these funds. Agencies are directed by these federal acts in how they manage their natural resources.

Environmental Compliance

A second set of laws that influence conservation planning falls under the "environmental compliance" umbrella. Two landmark pieces of legislation are the National Environmental Policy Act (NEPA) of 1969 and the Endangered Species Act (ESA) of 1973. NEPA was passed during the environmental movement of the 1960s in response to several environmental concerns ranging from clean water and air and pesticide contamination to declining wildlife species. The act declares a national policy "to encourage productive and enjoyable harmony between man and the environment and promote efforts to better understand and prevent damage to ecological systems and natural resources important to the nation" (42 USC § 4321).

The basic premise of NEPA is that federal agencies are required to prepare a detailed *environmental impact statement* (EIS) for any major federal action significantly affecting the environment. The act also establishes the Council on Environmental Quality to review government policies and programs for conformity with NEPA. Under NEPA, federal agencies are required to include an EIS in any proposed project or other major federal action that would potentially adversely impact the environment. An EIS includes an examination of environmental impacts and alternatives available to the proposed action. Prior to preparing an EIS, the agency must consult with other federal agencies having expertise on any environmental impact involved (e.g., US Fish and Wildlife Service concerning impacts to endangered species). Conservation planning conducted by federal agencies typically includes an EIS as part of their management plan.

The ESA provides protection to plants and animals considered threatened with or in danger of extinction. The authority for implementing and executing the ESA is delegated to the US Fish and Wildlife Service (FWS) for plants and animals or the National Marine Fisheries Service (NMFS) for marine life and anadromous fish. FWS and NMFS are responsible for determining which species are listed as threatened or endangered and delineating critical habitats necessary for their survival. ESA includes subdivisions that provide further authority to FWS and NMFS. For example, the ESA declares that all federal departments and agencies must seek to conserve endangered and threatened species and to "utilize their authorities in furtherance of the purposes of this Act" (16 USC. 1531–1544; Section 2). These sections include determining or listing a species threatened or endangered (Section 4), prohibited actions for listed species (Section 9), and penalties and enforcement procedures (Section 11).

Section 7 of ESA outlines procedures for interagency cooperation to conserve federally listed species and designated critical habitats. It requires federal agencies to consult with FWS to ensure that they are not undertaking, funding, permitting, or authorizing actions likely to jeopardize the continued existence of listed species or destroy or adversely modify designated critical habitat. It outlines processes for consultation to address listed species or species proposed for listing or proposed critical habitat. Other paragraphs of Section 7 establish the requirement for federal agencies to initiate early consultation or for FWS and NMFS to prepare biological opinions and issue incidental *take permits*. "Take" includes the physical removal of an individual animal from the wild, killing the animal, removal of habitat so it can no longer be used by the species, and activities such as recreation or machinery noise that make an area unsuitable for habitation by the species. Take can be authorized for scientific purposes and for commercial development purposes if plans are approved that mitigate for that take (usually of habitat).

Section 10 also allows take of listed species under specified circumstances through the development of a Habitat Conservation Plan (HCP) or some other similar agreement. An HCP is most often used when a number of similar activities will occur over a broad area, for example, when city and county governments want to expand their zone of commercial and residential development or when a transmission line (and associated right-of-way corridor) will cross numerous governmental jurisdictions. Under Section 10 incidental take is authorized through a variety of voluntary agreements to conserve or minimize and mitigate impacts on fish and wildlife, including (1) Candidate Conservation Agreements, (2) Safe Harbor Agreements, and (3) HCPs with Implementation Agreements. Following is a brief description of each of these types of agreements and how they influence or are a part of conservation planning.

Candidate Conservation Agreements

FWS and NMFS offer landowners a policy option of entering into "prelisting" or Candidate Conservation Agreements (CCAs) that provide regulatory guarantees to those who voluntarily agree to protect habitat for fish and wildlife species before they are listed for protection under the ESA. Under a recently proposed FWS rule for CCAs, successful applicants will receive an *enhancement of survival permit* (ESP) if they agree to actions that provide a net benefit to specified unlisted species so that listing the species would be unnecessary if other landowners within the range of the species were to manage their land in the same fashion. If species covered by the CCA are eventually listed for protection, the ESP authorizes incidental take of those species by any action in accordance with the CCA.

Safe Harbor Agreements

Safe Harbor Agreements are voluntary arrangements between the FWS or NMFS and cooperating nonfederal landowners. Their purpose is to promote voluntary management for threatened and endangered species on nonfederal property while offering assurances to landowners of future regulatory restrictions in the management of these species on their land. The value of Safe Harbor Agreements is they allow landowners to provide a net conservation benefit and contribute to recovery of a listed species in exchange for the assurance that a return to the baseline habitat condition at the time of permit inception will not result in liability for unlawful take. For example, a landowner who signs a Safe Harbor Agreement has 5 breeding pairs of an endangered species. If the habitat is improved to increase the breeding pairs to 10, the landowner would not be penalized if this number were to return to the baseline in the future because of an action such as development of species habitat. As in CCAs, parties to approved Safe Harbor Agreements will receive an ESP that authorizes incidental take by actions consistent with the terms of the agreement.

Habitat Conservation Plans

HCPs outline specific steps that the applicant must take to minimize and mitigate impacts to the endangered species. Frequently these plans require habitat protection, restoration, and enhancement in one area in exchange for some lost habitat in another. FWS has developed a handbook to help guide applicants and planners through the HCP planning process (US Fish and Wildlife Service and National Marine Fisheries Service 1996). In exchange for an HCP, a landowner is authorized to incidentally take protected species by any action consistent with the plan. In addition, an applicant for an HCP may negotiate for long-term regulatory assurances that no additional mitigation will be required under the plan regardless of changes in circumstances over the life of the permit. HCPs can be prepared for a single or multiple species. They also can be prepared for unlisted species to provide landowners with coverage under the ESA should the species become listed in the future. Single-species HCPs are easier to develop and implement, but multispecies HCPs should be considered for areas containing more than one listed or candidate species and for long-term projects such as phased developments where there is potential for additional listings in the future.

Incentive Programs

A third set of laws that influence conservation planning are "incentive programs." In many areas of the country, the majority of lands are privately owned (fig. 5.1). Thus, conservation planning on private lands involves incentivizing private landowners in managing their natural resources to their benefit and the benefit of the public. Private lands provide for many ecosystem benefits, such as clean water or open spaces. Incentive programs in essence allow federal and state agencies to promote certain behaviors by private landowners. A landmark piece of legislation important in supporting conservation on private lands falls under "Farm Bill" programs. The Food, Conservation, and Energy Act of 2008, also known as the 2008 Farm Bill, is a $288 billion, five-year agricultural policy. The 2008 Farm Bill is a continuation of the 2002 Farm Bill and follows a long history of agricultural subsidy in energy, conservation, nutrition, and rural development. One of the three major components of the Farm Bill is providing baseline funding for conservation and working lands programs ($4 billion in 2008). Farm Bill programs encourage farmers and ranchers through financial and technical assistance to protect wildlife habitat, control soil erosion, and reduce polluted runoff.

The following examples of conservation programs funded by the 2008 Farm Bill support wildlife conservation on private lands. The majority of these programs are managed by the NRCS, which was established in 1935 with the mission of conserving natural resources on private lands. Originally established by Congress as the Soil Conservation Service, NRCS has since expanded to provide landowners technical and financial assistance to improve soil, water, air, plants, and animals that result in productive lands and healthy ecosystems. We provide a brief description of each program, in which participation is voluntary, and how funding and/or technical assistance is delivered.

Conservation Reserve Program

The Conservation Reserve Program (CRP) pays farmers annual rental payments under 10- to 15-year contracts to set aside lands of marginal production. CRP also provides up to 50% of costs in establishing conservation practices that address soil erosion, water quality, wet-

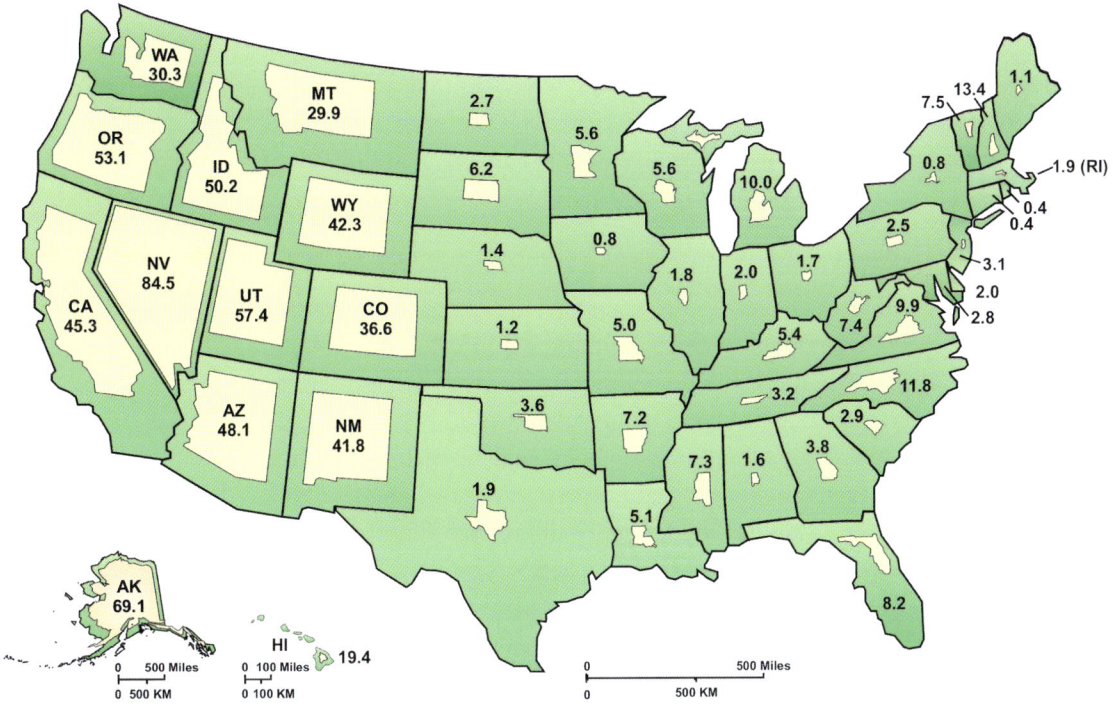

Figure 5.1. Federal land as a percentage of total state land area. Source: US General Services Administration, *Federal Real Property Profile 2004.*

land and forest enhancement, and wildlife management. Examples of these practices include establishing vegetative cover or trees on erodible cropland, planting native grasses, and placing buffer strips along stream banks to reduce pollution. In 2010, over 31 million acres were enrolled in the CRP program nationwide.

Grassland Reserve Program

The Grassland Reserve Program (GRP) enables landowners to restore or protect native grasslands on portions of their property through long-term or permanent easements. Maintaining and restoring native grasslands provide important wildlife habitat in addition to hunting and other recreational land uses. By 2010, approximately 3 million acres were enrolled in the GRP program nationwide.

Wetland Reserve Program

The Wetland Reserve Program (WRP) provides farmers the opportunity to restore, maintain, and protect wetlands on their property. Farmers can either sign up for a 10-year cost-share agreement or 30-year or permanent conservation easements with restoration cost-share funding. Most lands restored under WRP are marginal, high-risk, flood-prone lands. The WRP enables landowners to take these lands out of production and restore them to beneficial use such as wetland habitat for wildlife populations. By 2010, approximately 2 million acres were enrolled in WRP nationwide.

Wildlife Habitat Incentive Program

The Wildlife Habitat Incentive Program (WHIP) pays up to 75% of the cost to private landowners for enhancing wildlife habitat on their land. The program is not limited to agricultural lands but is open to any private landowners who would like to create wildlife-friendly habitat enhancements to a portion of their land, such as restoring native prairie grasses, performing forest management practices, or improving aquatic areas. In 2010, approximately $51 million was allocated to the states for WHIP projects on about 1 million acres of land.

Conservation Stewardship Program

The Conservation Stewardship Program (CSP) differs from the other programs by rewarding farmers and ranchers for environmentally friendly measures. The program offers payments to producers who maintain a high level of conservation on their land and agree to adopt higher levels of stewardship. Eligible lands include cropland, pastureland, rangeland, and nonindustrial forestland.

Environmental Quality Incentives Program

Similar to CSP, the Environmental Quality Incentives Program (EQIP) provides technical assistance, incentive payments, and cost sharing to farmers and ranchers to implement conservation practices on their lands. Contracts can be up to 10 years in duration, and allowable practices are based on a set of national priorities adapted to each state. These priorities range from reduction of point and nonpoint source pollution to watersheds and groundwater to the improvement of wildlife habitat for at-risk species.

State and Federal Conservation Planning

We now review some examples of the types of conservation planning conducted by state and federal natural resource agencies. Though the conservation planning is specific to each agency, there are some common general approaches or aspects (table 5.4). This review emphasizes background of the organization, basic processes followed in the development of that organization's management plan, and the way funding is tied to support management plan activities.

State Wildlife Action Plans

US laws and policies place the primary responsibility for wildlife management with the appropriate wildlife or natural resource agency in each state (Freyfogle and Goble 2009). State fish and wildlife agencies have a long history and success in conserving game species through support of hunter and angler license fees and federal excise taxes, such as through the Pittman-Robertson Act. How are species that are not hunted or fished (nearly 90%) managed on state and private lands? Obviously, there is a gap in wildlife conservation funding and the management of these other nongame species. Two funding programs address this gap in funding for the management of nongame and non-listed species: (1) State Wildlife Grants Program and (2) Landowner Incentive Program.

The *State Wildlife Grants Program* (created by Congress in 2001) provides federal funding to states to support projects that prevent wildlife from declining to the point of being endangered. Some examples of projects supported by this program are the restoration of degraded habitat, reintroductions of native wildlife populations, development of conservation partnerships with private landowners, and collection of data on declining species or species of concern. To make the best use of program funding, Congress charged each state and territory with developing a *Statewide Wildlife Action Plan*. These plans, technically referred to as Comprehensive Wildlife Conservation Strategies, assess the health of each state's wildlife and habitats, identify threats, and outline the actions needed to conserve them over the long term. Like the Pittman-Robertson funds, funds appropriated under the State Wildlife Grants Program are allocated based on a predetermined formula that evaluates the state's size and population. A nonfederal match requirement assures local ownership and leverages state and private funds to support conservation in each state. A specific format for the State Wildlife Action Plan comprising eight sections is required for funding support (see table 5.4).

The *Landowner Incentive Program* (LIP) is complementary to the State Wildlife Grants Program. The LIP was designed to benefit at-risk wildlife species and the habitats important to their survival. The FWS defines at-risk species as a "species of greatest conservation need" (high priority) in a State Wildlife Action Plan. LIP funds are allocated by Congress to FWS for distribution to state fish and wildlife agencies. Program funding can be used to develop and administer dedicated

Table 5.4. Overview of formal plans for federal and state agencies

State wildlife action plans	Statewide forest resource assessments	USFWS CCP	NRCS conservation plan	USACE land use plans
1. Distribution and abundance of wildlife (particularly low and declining populations)	1. Description of the priority landscape area and issues	1. Background (introduction, purpose, need for plan)	1. Collection and analysis	1. Specification of the water and related land resource problems and opportunities (relevant to the planning setting) associated with the federal objective and specific state and local concerns
2. Habitats and community descriptions essential to species conservation	2. Glossary of terms and acronyms	2. Refuge overview (location and size, physical resources, biological resources, socioeconomic environment)	a. Step 1: Identify problems and analysis	2. Inventory, forecast, and analysis of water and related land resource conditions within the planning area relevant to the identified problems and opportunities
3. Review of problems and factors adversely affecting species	3. Investing resources	3. Plan development (public involvement, planning process, review and revision)	b. Step 2: Determine objectives	3. Formulation of alternative plans
4. Proposed conservation actions and priorities for identifying species and their habitats	4. List of other plans consulted	4. Management direction (vision, goals, objectives, strategies)	c. Step 3: Inventory resources	4. Evaluation of the effects of the alternative plans
5. Monitoring plans for species and their habitats	5. Listing and description of stakeholder involvement	5. Plan implementation (proposed projects, funding personnel, monitoring and adaptive management)	d. Step 4: Analyze resource data 2. Decision support	5. Comparison of alternative plans
6. Procedures for plan reassessment and evaluation	6. Monitoring and reporting	6. Environmental assessment (NEPA)		6. Selection of a recommended plan based on the comparison of alternative plans
7. Coordination of plans with other agencies	7. Protocol for translating strategies into actions	7. Appendices	a. Step 5: Formulate alternatives	
8. Public participation plans	8. Strategies to address the priority landscape areas and issues		b. Step 6: Evaluate alternatives c. Step 7: Make decisions	
			3. Application and evaluation	
			a. Step 8: Implement the plan b. Step 9: Evaluate the plan	

BLM RMP	NPS GMP	USFS LRMP	DoD INRMP
1. Introduction	1. Identify relevant laws	1. Introduction	1. Description of the installation, its history, and its current mission
2. Purpose statement	2. Identify issues and concerns (scoping)	2. Desired future conditions, goals, and objectives (forestwide)	2. Management goals and associated time frames
3. Authority	3. Collect data	3. Standards and guidelines	3. Projects to be implemented and estimated costs
4. Organization and scope of an RMP document	4. Identify alternatives	4. Desired future conditions, goals, and objectives (management area)	4. Review of military mission and training requirements supported
5. Project history	5. Prepare draft plan	5. Monitoring, evaluation, research, and implementation	5. Legal requirements and biological needs
6. Location/setting	6. Revise and consult	6. Appendix and glossary	6. Role of the installation's natural resources in the context of the surrounding ecosystem
7. Overview of public involvement efforts	7. Approve final plan		7. Input from the FWS, state fish and wildlife agency, and the general public
8. Overview of consultation efforts			
9. Management framework			
10. Planning process	8. Implement the plan		
11. Opportunities and constraints			
12. Issues and issue categories			
13. Existing resource inventory/existing condition			
14. Goals and objectives			
15. Desired future condition			
16. Management action(s)/direction(s)			
17. Implementation procedures (monitoring, standards and guides, and plan revision or amendment)			

"private lands habitat program" that provides professional, technical, and financial assistance to private landowners. The LIP includes two funding tiers: Tier One (noncompetitive) and Tier Two (nationally competitive). Under Tier One, each state may receive funding for eligible projects up to $200,000 annually. When funding is available, the program will rank Tier Two grants and award grants through a national competition. Funding for LIP is collected from revenues from the Outer Continental Shelf oil and gas royalties deposited into the Land and Water Conservation Fund Act of 1965. Program funds are awarded to fish and wildlife agencies based on a two-tiered award system.

Statewide Forest Resource Assessment and Strategy

The 2008 Farm Bill resulted in the amendment of the Cooperative Forestry Assistance Act (CFAA) of 1973. The CFAA authorizes the secretary of agriculture to establish a variety of federal, state, and local cooperative forest stewardship programs to protect and manage nonfederal forestlands. Examples of funding programs include the Forestry Incentives Program, Urban and Community Forestry Assistance, Rural Fire Prevention and Control, and direct assistance to state forestry agencies. The State and Private Forestry organization of the US Department of Agriculture (USDA) Forest Service manages the majority of these programs. To be eligible for program funding, states are required to complete a *Statewide Forest Resource Assessment and Strategy* plan. Like the State Wildlife Action Plans, the Statewide Forest Assessments provide an analysis of forest conditions and trends in the state and delineate priority rural and urban forest landscape areas. The statewide assessment outlines long-term plans for investing state, federal, and other resources in the most efficient manner. Each Statewide Forest Resource Assessment and Strategy comprises eight sections that can be used to receive funding support (see table 5.4).

Federal Planning Overview

Several federal agencies are responsible either directly or indirectly for the management of US natural resources. For each of these agencies, conservation planning occurs as mandated by agency authorities or directives in meeting their mission. Though each agency uses a different term to describe its management plan, all federal agencies are subject to general requirements dictated by federal laws that outline a general process in the development and implementation of an agency's management plan. In general, a federal agency begins its planning process by publishing in the *Federal Register* and in local newspapers a Notice of Intent (NOI) to prepare or revise their version of the management plan. The NOI invites the public to identify issues and directly submit comments to the agency to be considered during the planning process. This phase of the planning process is typically referred to as the "scoping" phase.

Based on the information gathered (e.g., species data, public input), the agency then prepares a reasonable set of alternatives for managing the proposed public resources within the planning area. Management plan alternatives are designed to address issues identified during scoping and to comply with applicable laws and agency policy guidance. One alternative typically identified in the management plan is the "no action" alternative, which maintains the current management direction. Approaches to the identification of the "preferred" management alternative may differ among federal agencies, but at some point a given set of alternatives that includes the preferred option is presented to the public. Public meetings, mass mailings, and other informal discussions are used to solicit input into the prepared draft management plan. A public review period of 60–90 days is usually conducted to receive final public input prior to accepting the final version of the management plan.

FWS Comprehensive Conservation Plan

The FWS mission is to conserve, protect, and enhance fish and wildlife and their habitats directly on refuge lands and indirectly on private lands. Major agency responsibilities include the management of migratory birds, endangered species, freshwater and anadromous fish, wetlands, and the national wildlife refuge (NWR) system. Established in 1903 when President Theodore Roosevelt designated Pelican Island as the first NWR, the system has grown to include more than 150 million acres comprising 552 refuges. Each refuge requires a management plan to direct activities on FWS-owned properties. The National Wildlife Refuge System Improvement Act of 1997 (Improvement Act) requires FWS to prepare a Comprehensive Conservation Plan (CCP) for each NWR by the year 2012.

The CCP is a 15-year refuge management plan that describes the desired future conditions and provides long-range guidance and management direction to achieve refuge goals (Public Law 105–57; October 9, 1997). The development of a refuge CCP generally takes about a year, not including the NEPA process. There are five basic steps to the CCP process (see table 5.4):

1. Scoping phase: Identify public concerns and issues regarding the refuge through public meetings and other forms of data collection.

2. Formulate plan: Outline key issues and concerns, long-term refuge goals, and viable plan alternatives to meet goals.
3. Write draft plan: Draft a plan identifying management alternatives and effects distributed for internal review. The draft plan is then released for public comment.
4. Revise plan: Revise the plan based on public comments and issue the final CCP.
5. Implement plan: Use the plan to guide refuge activities until next revision.

NRCS Conservation Plans

Since 70% of US land is privately owned (see fig. 5.1), providing support to private landowners is an important aspect of the NRCS mission. NRCS field agents provide conservation planning expertise and manage incentive programs under the 2008 Farm Bill (WHIP, CRP, EQIP, etc.). Development of NRCS management plans is outlined in the *National Biology Manual* (NBM), which contains policies and procedures for biological resource activities, and the *National Planning Procedures Handbook* (NPPH), which provides specific policies and procedures for management plan development. NRCS uses a three-phase, nine-step planning process (fig. 5.2). The planning process is dynamic and not necessarily conducted in a linear or chronological order. The NPPH planning process is as follows:

1. Collection and analysis
   Step 1: Identify problems and analysis, opportunities, and concerns.
   Step 2: Determine objectives.
   Step 3: Inventory resources and their condition and economic and social considerations related to the resources.
   Step 4: Analyze resource data gathered in planning step 3 to clearly define the natural resource conditions, along with economic and social issues related to the resources. This includes problems and opportunities.
2. Decision support
   Step 5: Formulate alternatives to achieve client's objectives, solve natural resource problems, and take advantage of opportunities to improve or protect resource conditions.
   Step 6: Evaluate alternatives to determine their effects in addressing the client's objectives and the natural resource problems and opportunities. Evaluate the projected effects on social, economic, and ecological concerns. Special attention must be given to those

ecological values protected by law (ESA) or Executive Order (wetlands). Consider alternatives that may provide economic returns to the client from managing or restoring aquatic and terrestrial habitats.

Step 7: Make decisions on desired alternative(s) and develop the conservation plan. Documentation needs to be developed showing that NEPA concerns have been addressed during the planning process.

3. Application and evaluation

Step 8: Implement the plan and client-selected alternative(s). The planner provides encouragement to the client for continued implementation. Financial assistance can be requested with plan on file.

Step 9: Evaluate the effectiveness of the plan as it is implemented and make adjustments as needed.

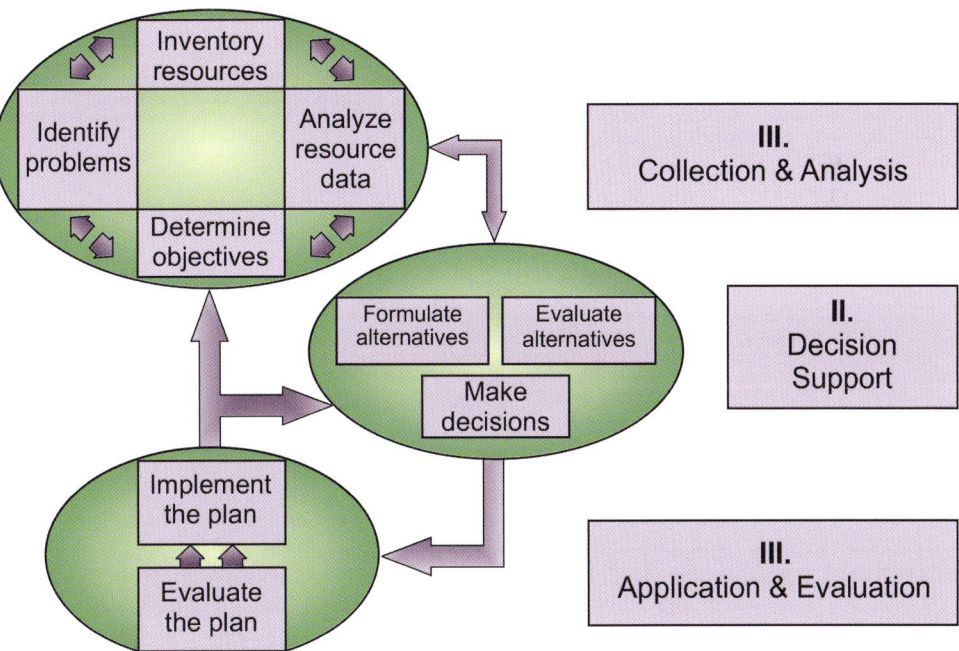

Figure 5.2. Example of the conservation planning process used by NRCS. Source: https://aglearn.usda.gov/customcontent/NRCS/Consplan/module3/3phase9step1.html.

US Army Corps of Engineers Land Use Plans

The US Army Corps of Engineers (USACE) is a federal agency and US Army command comprising civilian and military personnel. USACE is involved in a wide range of public works support to the nation and the Department of Defense throughout the world. The USACE mission is to provide public engineering services and strengthen the nation's security, energize the economy, and reduce risks from disasters. Some of its activities include planning, designing, building, and operating locks and dams; design and construction of flood protection systems; design and construction management of military facilities for the army and air force; and restoration of environmental regulations and eco-systems. USACE is, for example, the leading provider of outdoor recre-ation and hydropower capacity in the United States (US Army Corps of Engineers 2016). Through its Civil Works Program, USACE carries out a wide array of projects that provide coastal protection, flood protection, hydropower, navigable waters and ports, recreational opportunities, and water supply.

Water-conservation planning for USACE initially evolved from the Flood Control Acts (1928, 1936) and Rivers and Harbors Act (1925) to the more modern "308 Reports," comprehensive studies of US river basins. Water-conservation planning for USACE is articulated in the *USACE Principles and Guidelines* (P&G). The six-step planning process is described in the *P&G* as follows (see table 5.4):

1. Specification of the water and related land resource problems and opportunities (relevant to the planning setting) associated with the federal objective and specific state and local concerns
2. Inventory, forecast, and analysis of water and related land resource conditions within the planning area relevant to the identified prob-lems and opportunities
3. Formulation of alternative plans
4. Evaluation of the effects of the alternative plans
5. Comparison of alternative plans
6. Selection of a recommended plan based on the comparison of alter-native plans

Bureau of Land Management Resource Management Plans

The Bureau of Land Management (BLM) was created in 1946 by merg-ing the Grazing Service with the General Land Office. The agency's mis-sion is to sustain the health, diversity, and productivity of the nation's public lands for the use and enjoyment of present and future generations (US Bureau of Land Management 2011). The BLM manages over 261 million acres of public lands (most of these lands are in the west) and approximately 700 million acres of the subsurface minerals of both public and private lands. When the BLM was initially created, there were over 2,000 authorities for managing these public lands. As a result, Congress enacted the Federal Land Policy and Management Act of 1976 (FLPMA), which recognized the value of the remaining public lands by declaring BLM lands would remain in public ownership and gave the agency the mandate of "multiple use" management.

Under the FLPMA, BLM was also required to develop comprehensive land use plans called Resource Management Plans (RMPs). RMPs are developed for all BLM-owned lands, including resource areas, national monuments, and national conservation areas. Like those of other federal agencies, the BLM RMPs offer opportunities for public input through the NEPA process. There are 17 components of an RMP (see table 5.4):

1. Introduction
2. Purpose statement
3. Authority
4. Organization and scope of an RMP document
5. Project history
6. Location/setting
7. Overview of public involvement efforts
8. Overview of consultation efforts
9. Management framework
10. Planning process
11. Opportunities and constraints
12. Issues and issue categories
13. Existing resource inventory/existing condition
14. Goals and objectives
15. Desired future condition
16. Management action(s)/direction(s)
17. Implementation procedures (monitoring, standards and guides, and plan revision or amendment)

National Park Service General Management Planning

The National Park Service (NPS) is responsible for the management of national parks, many national monuments, and other conservation and historical properties. Established through the authority of the National Park Service Organic Act (1916), the NPS oversees 393 units

(58 designated as national parks) comprising approximately 84 million acres. The agency's mission is to preserve unimpaired the natural and cultural resources and values of the national park system for the enjoyment, education, and inspiration of this and future generations (16 U.S.C. § 1.1). The planning process for NPS stems from a base document called *Statement for Management* (SFM) that outlines procedures with the *General Management Planning* (GMP) document. NPS units rely on GMPs to direct management and development for 10–15 years before reevaluation. Each GMP is a collection of eight action plans that focus on the following key resources: wilderness, wildlife, history, archeology, paleontology, geology, recreation, and access. The NPS prepares GMPs using an eight-step process (see table 5.4):

1. Identify relevant laws
2. Identify issues and concerns (scoping)
3. Collect data
4. Identify alternatives
5. Prepare draft plan
6. Revise and consult
7. Approve final plan
8. Implement the plan

US Forest Service Land and Resource Management Plans

The US Forest Service (USFS) is an agency of the USDA that oversees 155 national forests and 20 national grasslands that collectively encompass approximately 193 million acres (US Forest Service 2011). Major divisions of the agency include the National Forest System, State and Private Forestry, and the Research and Development branch. The Forest Reserve Act (1891) originally authorized withdrawing land from the public domain as "forest reserves" managed by the Department of Interior. Later, the Transfer Act (1905) transferred the management of forest reserves from the Department of Interior to the Bureau of Forestry, which later became the USFS under USDA.

The National Forest Management Act (NFMA 1976, 1990) reorganized and expanded management of national forest lands. The NFMA requires the secretary of agriculture to assess forestlands and develop a resource management plan based on multiple-use, sustained-yield principles for each USFS unit. The Land and Resource Management Plans (LRMPs), also known as Forest Plans, are the product of a comprehensive notice and comment process established under NFMA. LRMPs provide direction for all future decisions in the planning area. The secretary must revise and update the management plans at least once every 15 years. The general outline for LRMPs comprise six sections (see table 5.4):

1. Introduction
2. Desired future conditions, goals, and objectives (forestwide)
3. Standards and guidelines
4. Desired future conditions, goals, and objectives (management area)
5. Monitoring, evaluation, research, and implementation
6. Appendix and glossary

Department of Defense Integrated Natural Resources Management Plans

The Department of Defense (DoD) is responsible for coordinating and supervising all agencies and functions related to national security and the US Armed Forces. The DoD comprises the Office of Secretary of Defense (OSD); the Departments of Army, Navy, and Air Force; and other support agencies. The DoD manages over 30 million acres of various ecosystems that represent the major land and climate types in which military personnel may be expected to fight wars. The primary function of DoD lands is to support the test and training mission for the US Armed Forces; however, DoD also is responsible for conserving and protecting biological resources under the authority of the Sikes Act (1960) with assistance from FWS.

In 1997, the Sikes Act was amended, requiring the military installations to develop and implement mutually agreed on Integrated Natural Resource Management Plans (INRMPs) through voluntary cooperative agreements between the DoD installation, FWS, and the respective state fish and wildlife agencies. INRMPs are planning documents that allow DoD installations to implement landscape-level management of their natural resources while coordinating with various stakeholders. INRMPs are reviewed and updated annually and reapproved (requiring the signatures of DoD, FWS, and the appropriate state agency) every five years. The INRMP process also takes into account military mission requirements, installation master planning, environmental planning, and outdoor recreation. These are the basic elements of an INRMP (see table 5.4):

1. Description of the installation, its history, and its current mission
2. Management goals and associated time frames
3. Projects to be implemented and estimated costs
4. Review of military mission and training requirements supported
5. Legal requirements and biological needs

6. Role of the installation's natural resources in the context of the surrounding ecosystem
7. Input from the FWS, state fish and wildlife agency, and the general public

General Characteristics of Planning

We began the chapter by presenting the value and reasons for conservation planning. In short, conservation planning serves to guide our management actions in a way to ensure the careful evaluation and execution of management plans, the basis for conservation planning. Next we reviewed some formal management plan outlines currently used by both state and federal agencies and the authorities through which these agencies implement conservation planning on private and public land. Through our review of these formal processes, we observed some common elements or themes to management plans. Now we emphasize a few of these common elements and propose a generic management plan model. Management plans are an integral component in managing wildlife habitat and their populations—a skill that wildlife practitioners should learn and refine in the practice and application of wildlife management.

The conservation planning process can be summarized as follows. Conservation planning begins with the notion that we are dissatisfied with the status quo; thus, conservation planning is problem driven. A declining endangered population, for example, may provide the reason for a wildlife manager's plans or actions. Because all environments are always changing, a manager also has to plan ahead even if currently satisfied. In either case, addressing a management problem or keeping things the way they are into the future requires some sort of image of a desired state (goals) and how to get there (objectives).

Thus, the conservation planning process begins with a situational diagnosis framed by the clear articulation of goals and objectives. Based on our evaluation of goals and objectives, we then begin to formulate predictions of likely outcomes from the suite of alternatives identified. Next, we conduct a feasibility analysis to determine what would we ideally desire and try to develop reasonable options for further consideration. We evaluate all possible options. Based on our findings, which include "reality" constraints such as budget and consultation with stakeholders, we select and implement a given set of alternatives in our plan. We repeat this process (conservation planning is an iterative process) because we realize that plans need to be adjusted and revisited from time to time.

Conservation planning is analogous to the scientific method learned in elementary school, which explained how to learn things. First, you observe a condition and form a hypothesis. You test your hypothesis in an experiment and compare the results to your hypothesis. You either confirm your hypothesis or repeat the process with a revised hypothesis. It was probably the first step-by-step, iterative, problem-solving approach that you experienced. Conservation planning is simply the scientific method dressed up, modified, and recycled. Consider, for example, the formal conservation planning process used by TNC (fig. 5.3). TNC's conservation planning process involves four steps related to their ecoregional assessments: (1) identify and define your project, (2) develop strategies and measures, (3) implement strategies and measures, and (4) use results to adjust and improve. Similarly, we suggest that conservation planning involves five steps: (1) state goals and objectives, (2) identify and assign tasks, (3) conduct tasks, (4) evaluate

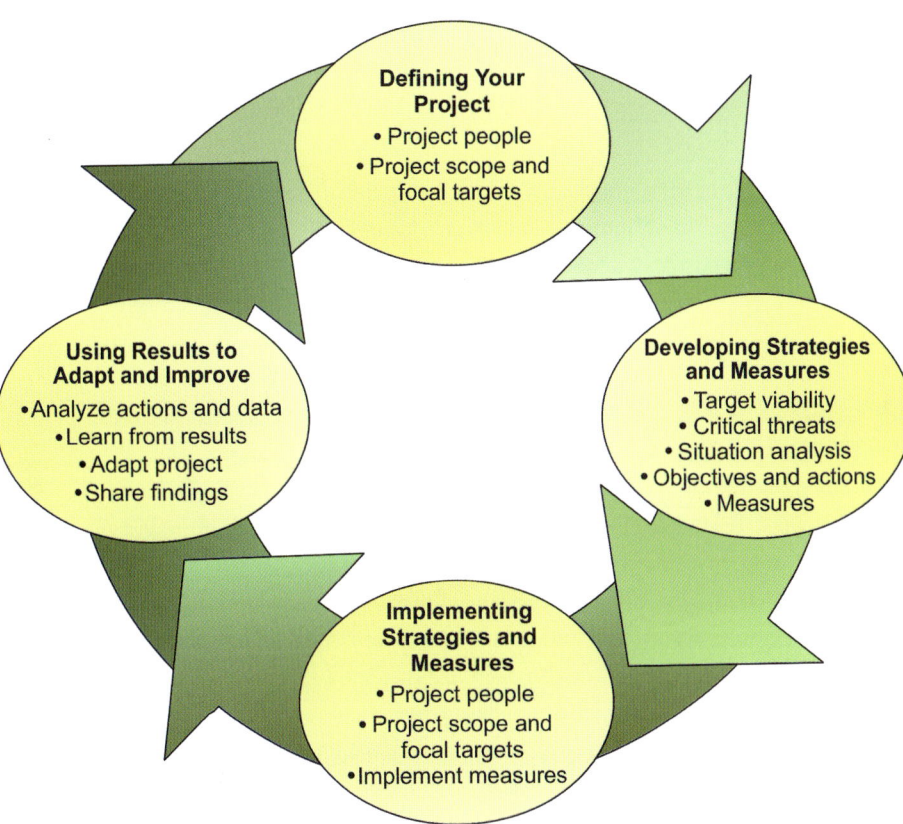

Figure 5.3. Example of the conservation planning process used by The Nature Conservancy. From The Nature Conservancy (2007).

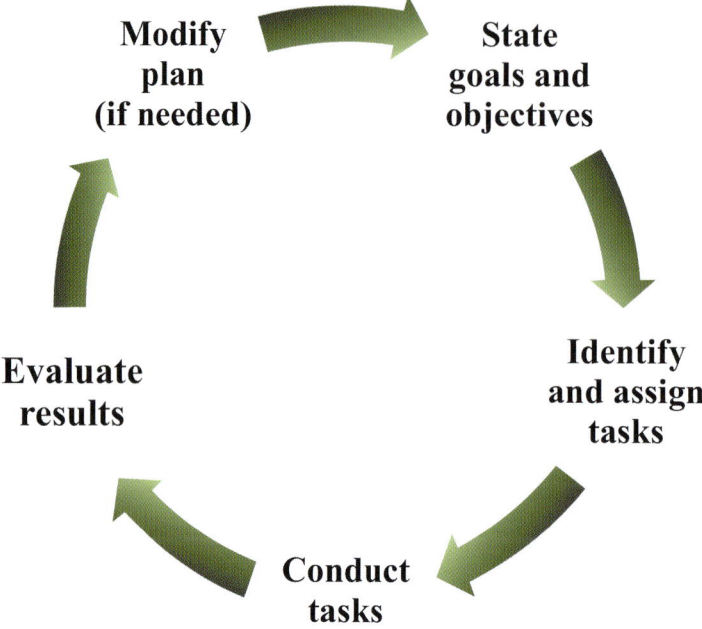

Figure 5.4. Five basic steps in conservation planning. Image by Roel R. Lopez.

results, and (5) modify the plan if needed (fig. 5.4). These five steps are used to frame the major components of management plans that we propose and review and can serve as the basis for conservation planning.

Management Plan Overview

From our review of formal plans for state and federal agencies (see table 5.4), we can generalize and define a generic approach in the development of a management plan. This review describes the basic components used in developing management plans that wildlife managers are likely to encounter throughout their careers: (1) introduction and purpose, (2) management analysis and alternative recommendations, (3) final recommendations and implementation, and (4) supporting documents. These sections of the plan are outlined in table 5.5, which supplements the following brief descriptions.

Introduction and Purpose

The first section of the management plan introduces the purpose of the plan. It answers the question "Why are we making this plan?" It describes the current land uses and specific ecological site descriptions for

Table 5.5. Example outline used in developing a landowner's wildlife management plan

Major sections	Plan elements	Description
Introduction and purpose	Purpose	Introduces the reader to the purpose of the plan. Answers the question "Why are we making this plan?"
	Owner information	Provides information about the property owner or client. Describes the purpose of the property from the landowner's perspective (recreational use, consumptive, etc.).
	Property description	Provides information about property location, including general locator and vegetation cover maps, legal description, area size, historic and current land uses, and other relevant biological descriptions (soils, topography, cover type, flora, fauna, etc.).
	Target species	Identifies species targeted for management. Normally, target species are stated within the landowner's objective.
	Goals and objectives	Describes the wildlife management goals and objectives for property. A landowner *goal* may include increasing the number of blue birds. A set of *objectives* used to address this goal may include identifying one or more viable management practices and recommending practice(s) to the landowner that would best achieve the goals.
Management analysis and alternative recommendations	Feasibility assessment	Reviews species requirements and environmental constraints that will frame or "bound" management recommendations. Describes feasible options for landowner consideration. A "management constraints" section may be included to identify and discuss any limitations to proposed management activities.
	Alternative recommendations	Outlines proposed management alternatives or options based on the feasibility assessment (species requirement, constraints, etc.) to support recommendations. A "no management" option is a viable alternative.
	Proposed target areas	Identifies target areas where the proposed management activities are to occur. A map is often included to aid in identifying those areas.
Final recommendations, budget, and time line	Final recommendations	Outlines final management recommendations to the landowner. Identifies how recommendation(s) will support and achieve management goal.
	Time line	Outlines a schedule of events of all proposed activities or necessary steps needed to implement the plan. Typically includes a table or some other calendar organized by practice/activity and when that activity should be conducted (installation of nest box, maintenance, monitoring, etc.).
	Budget	Provides a summary of costs for implementing the management plan. Typically includes a table with proposed number of acres treated and cost per treatment organized by management practice.
	Monitoring	Outlines a monitoring protocol, which is essential in the "feedback" loop process. Monitoring is essential to determine success and clearly define the metrics that will be monitored to know when success is achieved (e.g., successfully fledging quail).
Appendixes, glossary, and literature cited		Includes supporting information relevant to the wildlife management plan. This can include popular literature, information from websites, or scientific publications, e.g., technical assistance sources and contacts, material/equipment sources, or census and monitoring forms.

the planning. In describing current land uses, remote sensing such as aerial photography or other GIS data can be used to physically describe the planning area to include general vegetation cover, area of management units, soil types, property boundaries, hydrology, topography, transportation access (roads, trails, etc.), and other relevant spatial features. This review of the planning area identifies any site constraints prior to developing plan alternatives. Maps are often used at this point to "tell the story" and describe the current state of the area in question.

The description of the owner of the property and the owner's goals and objectives for the property are introduced in the first part of the management plan. A brief to extensive review of life history requirements for target wildlife species sometimes is included as background. This biological background is important, as it will later justify management recommendations based on the biology of the target species. Finally, the overall goal of the plan, as well as the more specific objectives that will lead to attaining the goal, is specified. In short, the first section of our generic management plan outline describes the current situation for further consideration by the wildlife practitioner.

Recommendations

The second section describes the process in developing and identifying feasible alternatives, an important part of any management plan. The wildlife practitioner, through this process, is made keenly aware of his or her "signature on the face of the land." Two primary factors will likely frame the development of management plan alternatives: the *biological needs* of the species (reference to the life history requirements); and the physical or *site constraints* of the planning area (both abiotic and biotic). The development of sound alternatives will have a profound effect on the quality of the final decision made in the management plan (Lichfield, Kettle, and Whitbread 1973).

A "good" plan cannot be chosen from a poor set of alternatives in conservation planning. During the initial stages of developing reasonable alternatives, the manager should develop a laundry list of possible actions. Alternatives come from people with varied backgrounds and experiences; however, not all alternatives are feasible. Thus, a second review of alternatives is typically done based on known constraints and available resources. Feasibility criteria can be based on cost-benefit analyses, anticipated social and/or political support, or other predetermined criteria. In addition to the review of feasible alternatives, this section of the management plan may also identify target areas for proposed management activities.

Implementation

The implementation section identifies final recommendations for the landowner to consider. In addition to reviewing specific management recommendations and supporting management actions or prescriptions, the management plan outlines the proposed time line of events or management plan milestones, costs or budget requirements for implementing the plan, management guidelines (BMPs), and monitoring and evaluation protocols. This section is important because it provides the details needed to successfully implement and evaluate the management plan. The monitoring and evaluation protocols must be included. Management plans are similar to roadmaps. You need to know when you will get there (reach your goals) based on evaluation criteria that are measurable and related to the goals and objectives of the management plan.

Supporting Documents

The fourth section concerns supporting materials or documents related to the main body of the plan. Though the term "supporting documents" gives a sense of unimportance, this section is important to the full understanding and implementation of the management plan. Appendixes, glossary of terms, and literature cited serve to "bookend" the management plan. In more formal management plans for federal agencies, a copy of an EIS or Biological Opinion may be included in this section for review by the reader. In the case of a management plan for a private landowner, popular literature, "how-to" publications, equipment lists and suppliers, building plans (e.g., nest boxes), sampling data forms, and other relevant planning documents are typically included.

Summary

I. What Is Planning?

    A. Conservation planning is a progression of steps in which we determine what we have, what we want, and how to get there.

    B. Definition: The deliberate social or organizational activity of developing an optimal strategy for solving problems and achieving a desired set of objectives.

    C. Important elements of planning include the following:

        1. Future control

        2. Problem solving

        3. Team effort

4. No single approach
5. Adaptive framework
6. Intention to implement

D. Establishment of goals and objectives is critical to successful management.

II. Types of Planning

A. There are many types of planning, including but not limited to these:
1. Land use plans
2. Transportation plans
3. Historic preservation plans

B. Collaboration across agencies and among resource managers is critical in the planning process.

III. Major Legislation and Policy

A. Conservation planning is typically driven by three basic frameworks:
1. Agency directives or authorities
2. Environmental compliance
3. Incentive programs

B. Management plans are often required in order to receive and use conservation funding.

IV. State and Federal Conservation Planning

A. Though conservation planning is specific to each agency, there are some general approaches or aspects that each has in common.

B. It is important that managers understand the background of major organizations, the basic processes that they follow in the development of management plans, and the way that funding supports management plan activities.

V. General Characteristics of Planning

A. Conservation planning guides management actions in a way to ensure the careful evaluation and execution of management plans.

B. Conservation plans have common elements:
1. The process begins with a situational diagnosis framed by the clear articulation of goals and objectives.
2. This is followed by a feasibility analysis where it is determined what is desired and reasonable.
3. Evaluation of all possible options based on findings that include "reality" constraints such as budget and in consultation with stakeholders is an iterative process.

VI. Management Plan Overview

   A. There are four major components of any management plan.

      1. Introduction and Purpose

         a. Introduces the reader to the purpose of the plan and answers the question "Why are we making this plan?"

         b. Describes the current land uses and specific ecological site descriptions for the planning area in question

      2. Recommendations

         a. Describes the process of developing and identifying feasible alternatives

         b. Generally framed by two factors:

            i. Biological need

            ii. Site constraints

      3. Implementation

         a. Reviews specific management recommendations and supporting management actions or prescriptions

         b. Outlines the proposed time line of events or management plan milestones, costs or budget requirements for implementing the plan, management guidelines (BMPs), and monitoring and evaluation protocols

      4. Supporting Documents

         a. Include appendixes, glossary, and literature cited sections

         b. Might include other information as required by individual agencies (e.g., EIS)

         c. Support and bookend the management plan

Literature Cited

Freyfogle, E. T., and D. D. Goble. 2009. *Wildlife law: A primer*. Washington, DC: Island Press.

Lichfield, N., P. Kettle, and M. Whitbread. 1973. *Evaluation in the planning process*. Oxford: Oxford University Press.

Lopez, R. R., B. Hayes, M. W. Wagner, S. L. Locke, R. A. McCleery, and N. J. Silvy. 2006. Integrating land conservation planning in the classroom. *Wildlife Society Bulletin*. 34:23–28.

The Nature Conservancy. 2007. *Conservation action planning handbook: Developing strategies, taking action and measuring success at any scale*. Arlington, VA: The Nature Conservancy.

US Army Corps of Engineers. 2016. http://www.usace.army.mil/Home.aspx.

US Bureau of Land Management. 2011. http://www.blm.gov/wo/st/en.html.

US Fish and Wildlife Service and National Marine Fisheries Service. 1996. *Habitat conservation planning and incidental take permit processing handbook.* November 4. http://www.nmfs.noaa.gov/pr/pdfs/laws/hcp_handbook.pdf.

US Forest Service. 2011. http://www.fs.fed.us/aboutus/.

Yoe, C. E., and K. D. Orth. 1996. *Planning manual.* Institute for Water Resources Report 96-R-21. Alexandria, VA: Institute for Water Resources.

6 Emerging Issues

A variety of emerging issues are changing and complicating wildlife habitat management in North America and around the world. As we discuss these issues, many will already be familiar to you, as they are common topics in the media, in politics, and within the wildlife profession. Global climate change, decline in water quality and availability, loss and fragmentation of habitat, broadening invasive species and diseases, increased human-wildlife conflicts, and urbanization and sprawl commonly appear on the evening news in some form or another. It should come as little surprise that many of the issues that humans are wrestling with politically and ecologically also have a large impact on wildlife habitat. We are part of these same ecological systems after all. For this reason we discuss both the large-scale ecological issues and the explicit social and political components. Now, more than ever, wildlife habitat management is both an ecological and a policy endeavor.

In this chapter, we present some of these important issues and case studies that demonstrate wildlife habitat management in action. This is not a full list of issues but represents some of the major challenges and factors seen in wildlife habitat management today. It is impossible to fully catalog every possible impact that these issues are likely to have. We instead briefly focus on these issues and how they impact some aspects of wildlife habitat management.

Global Climate Change

"Global climate change" and "global warming" are terms that most people in North America should be familiar with if not entirely sure of the definitions. "Global warming" is the warming of the earth due to the emission of heat-trapping greenhouse gases into the atmosphere (US Environmental Protection Agency 2016). "Global climate change" is the catchall term that describes long-term climate impacts, including alterations to precipitation and temperature patterns. These changes include new weather patterns, sea-level rise, melting glaciers, and many other biosphere alterations. Specific wildlife habitat impacts range widely from coral bleaching to changes in structure and location of terrestrial vegetation communities, wildlife range contractions, and alterations to soil chemistry. The media cover these issues extensively; politicians argue about them incessantly; and scientists, environmentalists, and others issue warnings about the expected impacts.

The conversation regarding global climate change is widespread, political, and often heated. However, scientists continue to document impacts of global climate change on wildlife habitat across the world. For instance, Adams-Hosking et al. (2011) used sophisticated modeling techniques to predict that koala (*Phascolarctos cinereus*) habitat would dramatically change in Australia due to global climate change. They projected a shift of habitat to the east and south in conjunction with an overall contraction of available range. Similarly, Galbraith et al. (2002) discussed the dramatic impacts of sea-level rise resulting from glacial melt on coastal wildlife communities in Northern California. They projected a dramatic 20%–70% loss of intertidal habitat in their study sites. Burns, Johnson, and Schmitz (2003) warned of the problems that static national parks and bioreserves face as global climate change alters the geographic distribution of wildlife habitats. They ultimately surmised that these immovable plots of land would fail to protect biodiversity in the face of overwhelming changes such as new species interactions, species extinctions, and structural changes in vegetation. Bellard et al. (2012) reviewed a variety of climate change research focused on biodiversity and found a vast majority predicted "alarming" declines in biodiversity worldwide.

What are some scenarios that wildlife habitat managers may have to respond to, and what strategies might they use? Responses to global climate change are varied based on situation. Responses range from wetland development and coastal barriers to help attenuate sea-level rise to species population modeling and habitat projections and policy changes to help legislate climate change mitigation. Thus, the predictive power of modeling has become incredibly important in planning responses to global climate change. For instance, Johnson et al. (2010) modeled the impacts of global climate change on prairie wetland complexes. Better known as prairie potholes, these are a mixture of permanent, temporary, and seasonal freshwater wetlands located throughout north-central North America and are critical for migratory waterfowl, flood control, improved water quality, and other ecosystem services. The authors found that these wetlands were likely to greatly suffer as a result of climate change. They predicted that the prairie pothole region will be too dry in the west and the east will have too few functional wetlands to support the current numbers of waterfowl and other wetland species.

Similarly, LaFever et al. (2007), modeled the impacts of sea-level rise on Lower Keys marsh rabbits (*Sylvilagus palustris hefneri*), an endangered marsh rabbit subspecies endemic to the Lower Florida Keys and dependent on grasslands located only at specific elevations above sea level. The Lower Florida Keys have a unique vegetative and freshwater structure highly dependent on elevation above sea level. Even small changes in sea level will have large impacts on this ecosystem. LaFever and his colleagues modeled vegetational changes likely to occur under different sea-level-rise scenarios. The challenge was to craft realistic management scenarios with limited vegetation and land availability under various sea-level situations. A reduction in marsh rabbit habitat is inevitable in the face of sea-level rise; however, managers, researchers, and policy makers were challenged to create broad-based conservation plans that include people and marsh rabbits, focus on adaptive management, and protect climatically important areas. In the case of the Lower Keys marsh rabbit, modeling efforts were used by managers in determining where to best invest resources (habitat restoration activities) in the face of climate change to increase the long-term outlook of the species' recovery. Management of wildlife habitat as global climate change impacts ecosystems all over the world will require similar decisive actions, insightful data collection, and modeling and policy changes.

Water Issues

Water is a basic need for humans and wildlife alike and is critical to many ecosystem processes. The increase in human population in North America, the growing offtake of water, and the alteration of hydrological cycles have dramatically altered water quality and availability in many ecosystems. Biodiversity and ecosystem health move to the forefront as current projections indicate that by 2025 humans will use more than 70% of accessible fresh water (Postel, Daily, and Ehrlich 1996). Almost all of the world's fresh water occurs underground, but groundwater pumping and pollution threaten these sources (Bouwer 2000). Agricultural operations and municipal water agencies rely heavily on pumped groundwater to meet current demands. However, as aquifers absorb more pollution and pumping exceeds aquifer recharge, groundwater sources require extensive management to maintain future water availability for humans and wildlife habitat. This is readily apparent in areas with high dependence on pumped groundwater such as the city of San Antonio, Texas (Edwards Aquifer), or farmers in the lower Midwest (Ogallala Aquifer). Communities in these areas struggle with balancing water availability with water need.

Additionally, surface waters such as rivers, streams, ponds, and lakes suffer from climate variability, pollution, channelization, invasive species, impoundments, and water offtake (e.g., irrigation, municipal water supplies). This places an emphasis on holistic water management plans that include both ground and surface waters and has led many cities and communities to strive for integrated water management (IWM) plans with strong stakeholder inclusion in the planning process (Bouwer 2000; Gulley 2015). Although Bouwer primarily discusses developing countries and their water needs, the concepts of IWM have been adopted in other water-scarce areas. IWM requires municipalities to balance water supply, quality, and demand (e.g., conservation, recycling). For instance, the Edwards Aquifer is an important groundwater source for major cities (e.g., San Antonio, Austin, San Marcos) in the Texas Hill Country. For nearly 50 years, water "wars" occurred between diverse stakeholders and culminated in the development of a Habitat Conservation Plan that balanced water demands with protection for eight listed species dependent on the Edwards Aquifer and associated Comal and San Marcos Springs. A plan was developed primarily through a consensus-based process of key stakeholder engagement, which allowed for the first time an IWM-like plan for the region (Gulley 2015).

Water quality and availability can also be negatively impacted by terrestrial vegetation management. For instance, *Tamarix* (salt cedar, tamarisk) has spread in riparian areas, floodplains, ditches, and wetlands throughout the southwestern United States. *Tamarix* is an invasive exotic shrub that colonizes water-rich areas and outcompetes many native vegetative species. Its high water intake and respiration rates combined with a tendency to grow into high-density monocultures leads to large amounts of water loss from colonized areas. Thus, wildlife habitat managers seek to reduce *Tamarix* abundance and restrict further spread (Shafroth et al. 2005).

They generally work to replant with native vegetation and maintain active programs to remove remnant *Tamarix*. However, research indicates mixed impacts from removal of *Tamarix* when variable water amounts are returned to the system. Additionally, wildlife occasionally struggles to recolonize when *Tamarix* is replaced with native vegetation. This places an additional onus on the manager to carefully weigh the costs and benefits of large-scale *Tamarix* removal on target lands.

Tamarix is often removed as a component of larger riparian area management efforts. Researchers and managers have long known that healthy forest buffers around riparian zones can filter pollutants from surrounding areas before they reach waterways (Anbumozhi, Radhakrishnan, and Yamaji 2005), which increases the value of *Tamarix* removal.

Similarly, wetlands can serve as filters from surrounding areas or as filters of the water itself. For instance, the Trinity River in Texas receives a great deal of pollution from the Dallas–Fort Worth (DFW) metropolitan area. As the river flows south of DFW, the water is directed through a series of specially created wetland cells located on a wildlife management area that helps clean out many of these pollutants. The Tarrant Regional Water District's George W. Shannon Wetlands Water Reuse Project was the first of its kind in the United States and is scheduled to become a functional water-supply alternative to the district's rapidly growing service area. The water district currently provides raw water to almost 2 million people in 11 counties. But that number is expected to increase to 4.3 million by 2060. Located on Texas Parks and Wildlife's Richland Creek Wildlife Management Area on the Navarro-Freestone County line, the system consists of a series of sedimentation ponds and wetland cells that naturally treat water diverted from the Trinity River and put it back into Richland-Chambers Reservoir for future use (Tarrant Regional Water District 2016). Innovative approaches that include habitat management strategies will address many of our future natural resource challenges, as illustrated by the DFW wetland reuse project.

Invasive Species and Disease Issues

Changing climates, urbanization, and other changes have facilitated both invasive species and changes in wildlife disease ecology. Managers are increasingly dealing with new paradigms caused by ecosystems impacted at fundamental levels. Thousands of species are emerging in new ecosystems and competing with native species, changing vegetative communities, impacting soil chemistry and water quality, and carrying native and exotic diseases. In this section, we briefly discuss the broad impacts of invasive species and disease ecology and a few management scenarios.

Invasive Species

Invasive species are an emerging and growing problem in wildlife habitat management. Many of the underlying issues we discuss, such as exotic diseases, water quality, and native species population declines, have roots in invasive species. We have already discussed invasive exotic *Tamarix* and the problems it poses for riparian ecosystems. The spread of *Tamarix* was caused by human actions. Invasive exotic species often follow humans to new areas. Invasive species cause direct and indirect competition, often support native and exotic diseases, interbreed with native species, and impact ecosystem quality and function (e.g., zebra mussels [*Dreissena polymorpha*], which lower water quality). For example, Pejchar and Mooney (2009) conducted a meta-analysis of invasive species literature that demonstrated invasive species severely and negatively impair ecosystem functions and impact ecosystem services. A meta-analysis conducted by Vilà et al. (2011) found that much of the invasive species research indicated that invasive vegetative species caused large impacts, such as decreases in local plant species and changes in nutrient cycling.

For example, feral hogs were introduced into the United States several hundred years ago and now inhabit large parts of North America. They prey on a variety of native species, serve as carriers for native and exotic diseases, damage agricultural operations, and contribute to declines in water quality (Mapston 2007; Parker et al. 2015). Researchers and managers have advocated aggressive population-control efforts, stakeholder collaborations, and education to reduce feral hog populations. Laws may encourage unregulated take, restrict importation, and fund outreach efforts to empower stakeholders on hog control. Despite these efforts feral hogs are likely a permanent fixture on the North American landscape.

Tamarix or other invasive exotics such as feral hogs, the Russian olive (*Elaeagnus angustifolia*), and the European starling (*Sturnus vulgaris*) have considerable impacts on wildlife populations and habitats and can be incredibly difficult to eliminate once established.

Diseases

Researchers and managers are documenting increasing numbers of emerging infectious diseases of wildlife and humans. This uptick is likely driven by multiple issues such as increases in human population density, human encroachment into wildlife habitat, urbanization, and changes in wildlife habitat such as that caused by global climate change (Daszak, Cunningham, and Hyatt 2000; Bradley and Altizer 2006; Harvell et al. 2009). Humans support disease transmission through movement and maintenance of domestic animals, introduction of free-ranging invasives, wildlife habitat destruction resulting in increased wildlife densities and contact with new diseases, and changes in disease ecology due to climate change and other habitat alterations. Many infectious diseases in humans come from exposure to zoonoses (diseases transmitted between animals and humans; Daszak, Cunningham, and Hyatt 2000) that are maintained in the environment by wildlife, such as Lyme disease and Chagas disease. As the ecosystem changes or moves (locations or elevations move as climates change), diseased wildlife may increasingly encounter people. Additionally, as these systems change, disease ecology may change to the point that wildlife increasingly encounters disease agents. For example, red abalone (*Haliotis rufescens*) was one of several abalone species in California to experience population declines as elevated seawater temperatures from global climate change encouraged transmission of withering syndrome (Harvell et al. 2009).

Atkinson and LaPointe (2009) described in wonderful detail the history and impacts of avian malaria and poxvirus on Hawaiian honeycreeper (Drepanidinae). The endemic Hawaiian honeycreeper is by most measures a success story highlighting the strengths of adaptive radiation because this group of birds filled a variety of habitats on the Hawaiian Islands. In a further testament to the Hawaiian Islands as a paradise on earth, mosquitoes were not present prior to European arrival on the islands. This changed as early sailors accidentally introduced mosquito eggs from sailing vessels onto the islands.

It is thought that avian malaria and poxvirus were introduced in the nineteenth centuries via nonnative passerines and domestic chickens, respectively. The mosquitoes effectively transmit both diseases. Both of these diseases have had population-level impacts on honeycreepers, with many species and subspecies going extinct or declining at alarming levels. Interestingly, the human introduction of a nonnative suite of parasitical and pathogenic species has combined with the impacts of global warming to exacerbate the problems. Currently, mosquitoes are limited by intolerable temperatures found at higher elevations. These low-risk zones are predicted to decline by 57% as temperatures warm and expose more honeycreepers to disease. Management of this onerous problem has been difficult. Strategies have included reduction of mosquito larval habitat, reduction of exotic wildlife that create larval habitat such as feral hogs, management of human-made water sources, consideration of vaccinations and other therapies, and preservation of low-elevation forests (Atkinson and LaPointe 2009). Much like feral hogs, however, this a problem that is unlikely to be solved but only managed.

Urbanization, Sprawl, and Habitat Impacts

As the human population has expanded, so too have human habitation needs. An overwhelming majority (84%) of people are expected to live in urban areas by 2050, with much of this transition occurring in developing nations (United Nations 2013). Agriculture and urban areas must expand substantially to meet these new demands, resulting in habitat fragmentation and loss. Much of this loss is projected to occur mainly in ecologically sensitive areas such as tropics and desert grasslands in developing countries (Fahrig 2003; Seto et al. 2011) but continues in North America as well. Ecological impacts are assured as agriculture expands and intensifies and urban areas multiply by converting less human-impacted rural areas to concrete cityscapes (Nelson 1992).

As expected, this has enormous potential impacts on wildlife habitats. In fact, many of our wildlife population problems are actually habitat and space problems. For example, mountain lions have a large home territory ranging from 100 to 400 square kilometers (39 to 154 sq mi) depending on variables such as sex and habitat type (Dickson and Beier 2002). Large space and habitat requirements make species such as mountain lions especially vulnerable to habitat loss and fragmentation. This is explicitly clear in areas with large and growing human populations such as Southern California, where managers have struggled to maintain corridors between remaining mountain lion habitat in the face of urban sprawl (Beier 1993; Morrison and Boyce 2008). Isolated mountain lion populations were found to have high risks of extinction without effective connectivity to other habitat. Morrison and Boyce cautioned that managers and conservationists should focus only on the most critical habitat needs when presented with habitat use competition with humans. They cautioned that ongoing investment, vigilance, realistic goals, agency commitment, and careful planning would be needed to continue successful mountain lion conservation. The difficulty is in navigating mountain lion needs even as habitat and habitat connectivity are threatened by continued human expansion.

As another example, the endangered Rio Grande silvery minnow (*Hybognathus amarus*) lives in the Rio Grande and Pecos River in New Mexico. Once inhabiting 3,862 kilometers (2,400 mi) of river, the Rio Grande silvery minnow now inhabits only 280 kilometers (174 mi), a 97% decline (US Fish and Wildlife Service 2007). Much of this decline is attributable to a variety of human causes, such as dewatering and water diversion, water impoundment, river channelization, nonnative species introduction, and water-quality decline. Water is a scarce and critically needed resource in the desert Southwest for both municipalities and agriculture. Management strategies have run the gamut, including increased research into minnow ecology, habitat restoration, captive breeding and reintroduction, implementation of adaptive management programs to encourage management changes when situations dictate, and engagement of stakeholders. The basic needs of the Rio Grande silvery minnow have placed it directly in the controversy, as it threatens to divert scarce water from other human needs, leading to a variety of lawsuits by environmental groups to ensure preferred conservation measures. The value of stakeholder engagement is high as all parties attempt to share an increasingly rare resource and avoid lawsuits in the future. However, this competition for limited resources is increasingly common because human-wildlife conflicts present onerous problems for managers and policy makers.

These conflicts are evident in social debates such as those surrounding the Endangered Species Act and other conservation laws, the role of the public trust versus private property rights, agricultural and wildlife tax valuations, human encroachment on wilderness areas, and Department of Defense (DoD) installations. These problems have prompted a variety of novel management collaborations that attempt to benefit both wildlife and stakeholders. For example, the red-cockaded woodpecker is a federally listed endangered species inhabiting the open pine forests of the southeastern United States from Virginia and west to Texas. Its primary habitat—the longleaf pine ecosystem—has been reduced since European settlement to approximately 3% of its original expanse (Texas A&M Institute of Renewable Natural Resources 2016).

Most recovery and management efforts for the red-cockaded woodpecker in eastern North Carolina are located on public lands, including DoD lands. Recovery efforts on DoD lands often limit the military's ability to train and maintain mission readiness. In order to meet recovery goals for the woodpecker and increase training flexibility for DoD, management efforts focused on private lands are critically important. This has resulted in innovative partnerships between the DoD, universities, and research institutions to engage private landowners in conservation efforts on private lands that include economic incentives to manage habitat for the red-cockaded woodpecker. These types of complex engagements with private landowners are representative of the efforts needed to conserve species and species habitats in the future.

Conclusion

A number of issues are emerging in wildlife habitat management. We discuss just a few of the important and broad-ranging issues, including global climate change, water quality and availability declines, habitat loss and fragmentation, invasive species and diseases, human-wildlife conflicts, and urbanization and sprawl. Society in general and wildlife habitat managers in particular must deal with these issues. Wildlife policy is often the way society chooses to confront these issues. On a societal level that includes everything from local to global communities, natural resource policies are debated, created, altered, and rescinded rather frequently. Some, such as the Endangered Species Act, are codified into law and become strong tools for species and habitat protection. Wildlife habitat managers play many roles: citizen of society, provider of information to policy makers, and implementer of management policies. It is therefore important that we mention both the ecology and the sociopolitical components intrinsic to wildlife habitat management. It is impossible to do justice to all issues, but every manager is likely to frequently encounter these select few.

Summary

I. Emerging Issues

A. Emerging issues are challenging wildlife habitat managers in North America and around the world.

B. These issues are broad, complicated, and fraught with difficulty.

C. We present these emerging issues with some accompanying examples that demonstrate wildlife habitat management realities and responses.

II. Global Climate Change

A. Global warming is the warming of the earth largely attributed to the emission of heat-trapping greenhouse gases into the atmosphere.

B. Global climate change describes the various climate and weather impacts resulting from global warming.

C. Global climate change is having demonstrable impacts on wildlife habitat around the globe.

D. Managers and policy makers have a variety of responses at their disposal; however, the actions will often be situational with extensive stakeholder involvement expected.

III.   Water Issues

A. Increasing human populations and individual water consumption have caused more water offtake and pollution.

B. These realities have necessitated more water use and development planning such as integrated water management.

C. Collaborative approaches including all stakeholders are important in crafting durable and effective water management plans.

D. The Edwards Aquifer process is an example of an effective collaborative approach to stakeholder engagement in water resource planning.

IV.   Invasive Species

A. Many underlying issues we discuss have roots in invasive species such as exotic diseases, water-quality issues, and native species population declines.

B. Invasive species cause direct and indirect competition, often support native and exotic diseases, interbreed with native species, and impact ecosystem quality and function.

C. This problem has become more acute as humans have increased in population size and expanded in impact.

D. Many researchers and managers advocate extensive invasive species prevention and removal (if practical) measures.

V.   Diseases

A. Increasing numbers of infectious diseases are emerging throughout the world.

B. This is likely caused by issues such as increasing human population density, human encroachment into wild areas, urbanization, and global climate change.

C. Humans further support disease transmission through actions such as domestic animal transport, introduction of invasive species, and alteration of native wildlife habitats.

VI.   Urbanization, Sprawl, and Habitat Impacts

A. Agriculture and urban areas must expand substantially to meet these new demands, resulting in habitat fragmentation and loss.

B. Much of this loss is projected to occur mainly in ecologically sensitive areas such as tropics and desert grasslands in developing countries but continues in North America as well.

C. Increased human-wildlife conflicts are a reality as humans expand into rural and wild areas.

D. Human-wildlife conflicts are another impetus for better stakeholder engagement processes.

Literature Cited

Adams-Hosking, C., H. S. Grantham, J. R. Rhodes, C. McAlpine, and P. T. Moss. 2011. Modelling climate-change-induced shifts in the distribution of koala. *Wildlife Research* 38:122–30.

Anbumozhi, V., J. Radhakrishnan, and E. Yamaji. 2005. Impacts of riparian buffer zones on water quality and associated management considerations. *Ecological Engineering* 24:517–23.

Atkinson, C. T., and D. A. LaPointe. 2009. Introduced avian diseases, climate change, and the future of Hawaiian honeycreepers. *Journal of Avian Medicine and Surgery* 23:53–63.

Beier, P. 1993. Determining minimum habitat areas and habitat corridors for cougars. *Conservation Biology* 7:94–108.

Bellard, C., C. Bertelsmeier, P. Leadley, W. Thuiller, and F. Courchamp. 2012. Impacts of climate change on the future of biodiversity. *Ecology Letters* 15:365–77.

Bouwer, H. 2000. Integrated water management: Emerging issues and challenges. *Agricultural Water Development* 45:217–28.

Bradley, C. A., and S. Altizer. 2006. Urbanization and the ecology of wildlife diseases. *Trends in Ecology and Evolution* 22:95–102.

Burns, C. E., K. M. Johnson, and O. J. Schmitz. 2003. Global climate change and mammalian species diversity in US national parks. *Proceedings of the National Academy of Sciences* 100:11474–77.

Daszak, P., A. A. Cunningham, and A. D. Hyatt. 2000. Emerging infectious diseases of wildlife—threats to biodiversity and human health. *Science* 287:443–49.

Dickson, B. G., and P. Beier. 2002. Home-range and habitat selection by adult cougars in southern California. *Journal of Wildlife Management* 66:1235–45.

Fahrig, L. 2003. Effects of habitat fragmentation on biodiversity. *Annual Review of Ecology, Evolution, and Systematics* 34:487–515.

Galbraith, H., R. Jones, R. Park, J. Clough, S. Herrod-Julius, B. Harrington, and G. Page. 2002. Global climate change and sea level rise: Potential losses of intertidal habitat for shorebirds. *Waterbirds* 25:173–83.

Gulley, R. L. 2015. *Heads above water: The inside story of the Edwards Aquifer Recovery Implementation Program*. College Station: Texas A&M University Press.

Harvell, D., S. Altizer, I. M. Cattadori, L. Harrington, and E. Weil. 2009. Climate change and wildlife diseases: When does the host matter the most? *Ecology* 90:912–20.

Johnson, W. C., B. Werner, G. R. Guntenspergen, R. A. Voldseth, B. Millett, D. E. Naugle, M. Tulbure, R. W. H. Carroll, J. Tracy, and C. Olawsky. 2010. Prairie wetland complexes as landscape functional units in a changing climate. *BioScience* 60:128–40.

LaFever, D. H., R. R. Lopez, R. A. Feagin, and N. J. Silvy. 2007. Predicting the impacts of future sea-level rise on an endangered lagomorph. *Environmental Management* 40:430–37.

Mapston, M. E. 2007. *Feral hogs in Texas*. AgriLife Extension B-6149, 03–07. College Station: Texas A&M University.

Morrison, S. A., and W. M. Boyce. 2008. Conserving connectivity: Some lessons from mountain lions in Southern California. *Conservation Biology* 23:275–85.

Nelson, A. C. 1992. Preserving prime farmland in the face of urbanization: Lessons from Oregon. *Journal of the American Planning Association* 58:467–88.

Parker, I. D., R. R. Lopez, R. Karthikeyan, N. J. Silvy, D. S. Davis, and J. C. Cathey. 2015. A model for assessing mammal contribution of *Escherichia coli*. *Wildlife Research* 42:217–22.

Pejchar, L., and H. A. Mooney. 2009. Invasive species, ecosystem services and human well-being. *Trends in Ecology and Evolution* 24:497–504.

Postel, S. L., G. C. Daily, and P. R. Ehrlich. 1996. Human appropriation of renewable fresh water. *Science* 271:785–88.

Seto, K. C., M. Fragkias, B. Güneralp, and M. K. Reilly. 2011. A meta-analysis of global urban land expansion. *PLoS One* 6:1–9.

Shafroth, P. B., J. R. Cleverly, T. L. Dudley, J. P. Taylor, C. Van Riper III, E. P. Weeks, and J. N. Stuart. 2005. Control of *Tamarix* in the western United States: Implications for water salvage, wildlife use, and riparian restoration. *Environmental Management* 35:231–46.

Tarrant Regional Water District (TRWD). 2016. Wetlands: Overview. http://www.trwd.com/wetlands.

Texas A&M Institute of Renewable Natural Resources. 2016. Assessing recovery opportunities for red-cockaded woodpeckers on private lands in eastern North Carolina. http://irnr.tamu.edu/media/233190/assessing_rcw.pdf.

United Nations. 2013. *World population prospects: The 2012 revision*. Department of Economic and Social Affairs: Population Division. ESA/P/WP.228. http://esa.un.org/unpd/wpp/publications/Files/WPP2012_HIGHLIGHTS.pdf.

US Environmental Protection Agency (USEPA). 2016. Climate change: Basic information. http://www3.epa.gov/climatechange/basics/.

US Fish and Wildlife Service. 2007. *Draft revised Rio Grande silvery minnow* (Hybognathus amarus) *recovery plan*. Albuquerque, NM: US Fish and Wildlife Service, Southwest Region.

Vilà, M., J. L. Espinar, M. Hejda, P. E. Hulme, V. Jarošík, J. L. Maron, J. Pergl, U. Schaffner, Y. Sun, and P. Pyšek. 2011. Ecological impacts of invasive alien plants: A meta-analysis of their effects on species, communities and ecosystems. *Ecology Letters* 14:702–8.

Index

Page numbers in *italics* refer to tables and figures; page numbers in **bold** refer to definitions.

conservation planning (*cont*)
agreements, voluntary; incentive
programs, private lands
Conservation Reserve Program
(CRP),176–77, 184
Conservation Stewardship Program
(CSP), 178, 184
Cooperative Forestry Assistance Act
(CFAA), 182
coordinate systems: Cartesian, **28**–*29*;
latitude and longitude, **29**; State
Plane, **29**; Universal Transverse
Mercator (UTM), **29**
coordinates, rectangular, 28–30
corridor, *12*, 14, 148, 206
cow as management tool. *See* grazing
management

Darwin, Charles, 2
data analysis, graphics and descriptive
statistics figures for, 66–69
delayed aerial ignition devices
(DAIDs), 95, 97
density: absolute, **39**; estimates of, 33,
49, 51; mean, 36; relative, **22, 39**, 42,
50; animal, **22**, 39–43, 45–52, 55, 120;
vegetation, **21**, 33–34, 36
Department of Agriculture, U.S., 182,
188
Department of Defense, U.S., *170–71*,
186, 189, 207
Department of Interior, U.S., *170–72*,
188
Department of Transportation, U.S.,
170–71
depth-of-edge influence zone, **16**
dimension analysis, 36
Dingell-Johnson Act, 172
direction: errors in, 27–28; values of, 26
disease, 85, 147, 149, 204–205. *See also
specific diseases*
distance sampling, 49–51
disturbance: ecology, **8**–9; human
induced, 8, 148–49; natural 8;
regimes, 76, 81; types of, *9*–11; succession, 16–17
dot grid, 30
double meridian distance (DMD)
method, 29
double parallel determination (DPD)
method, 29

drive count, **40**
drones, 53–54
Duck Stamp Act, 82

Eberhardt's removal method, 53
ecoclines, **11**, 14
ecological continuity, **12**
ecology: disturbance, **8**; landscape,
6–8
ecoregion, 6, 7, 167, *169*, 191
ecosystem services, 80, 82–83, 148
ecotones, **11**, 14
edge, **14**–*15*
edge effect, *12*, 15
edge contrast, **12**,
electronic distance measurement
device (EDM), 26
elevation, determining, 31
elms (*Ulmus* spp.), 37
endangered species, 44, 118, **128**,
148–49, 155, 173–75, 207
Endangered Species Act, 82, 153,
172–74, 207
endemic species, 44, 82, 155, 201, 205
enhancement of survival permit
(ESP), 175
environmental impact statement
(EIS), 173
Environmental Quality Incentives
Program (EQIP), 178, 184
evapotranspiration, **142**
evolution, theory of, 2
expeditions, history of, 1–2
experimental design, 23
experimental units, **23**
exploited populations, sampling,
51–53

Farm Bill, 169, 176, 182, 184
farmlands, **78**–80
federal agencies. *See specific agencies*
Federal Aid in Wildlife Restoration
Act, 172
Federal Land Policy and Management
Act (FLPMA), 187
Federal Register, 153, 182
Federal Water Pollution Control Act, 82
feral hogs (*Sus scrofa*), 142, 204
field test, preliminary, 25
figures, development of statistics,
70–71

home range, 5, 206
hunting: license, 132–33; regulations, 131; revenue, 131, *170–71, 172*, 179; seasons, 131; quotas, 130, 132. *See also* harvest management
hydrology, 7–8, 81, 142, 194

imagery, aerial, 22, 32, 56
imagery, satellite, 32, 53–54, 56
impact assessment, 66
incentive programs, private lands, 176–781
indices, animal, 40–43
indices, fruit, 39
Integrated Natural Resource Management Plan (INRMP), 189–90
integrated water management plan (IWMP), 202
invasive species: introduction and invasion of, 115, 118, 203–204; control, 85–86, 122, 136
island biogeography, theory of, 9–10
isogons, **27**

Jolly-Seber method, 46

deer, Florida Key (*Odocoileus virginianus clavium*), 44

Land and Resource Management Plan (LRMP), 188–89
Land and Water Conservation Fund Act, 182
land measurement: area determination, 30; linear distance, 26; units of measure, 25
land stewardship. *See* incentive programs, private lands
Landsat, 32
Landowner Incentive Program (LIP), 179–81
landscape ecology, **6–8**
laws and policies, U.S. *See specific laws and policies*
Leopold, Aldo: tools of management, 75, 84, 100, 118, 135; quote by, 164
lesser prairie-chicken (*Tympanuchus pallidicinctus*), 77
LiDAR (light detection and ranging), 32, 57
Lincoln-Peterson estimator, 45–46

line transects, 50
line-intercept method, 34–35
Linnaeus, Carl, 1
listed species. *See* endangered species; threatened species
longitude, **26**
Lyme's disease, 149, 205

macrohabitat, **4**
magnetic north, **26–27**
management plans: federal, 182–90; management plan example, 192–95, *193*; state, 178–82. *See also* conservation planning
map-scale ratios, **8.** *See also* scale
mark-recapture techniques, 45–47
marsh rabbit, Lower Keys (*Sylvilagus palustris hefneri*), 70, 201
mast, fruit, 38–39
maples (*Acer* spp.), 38
material data safety sheets (MSDSs), 136
maximum daily movement, 46
maximum sustained yield, 130
mean, **71**
mechanical treatments: benefits of, 81, 101, 112–13; checklist *101–11*; cost considerations, 108, *111*, 114; forest regeneration methods, 115–17; monitoring and assessment of, 109, 117; planning for and management plans, 107–108, 113–14; overview, 100; tools and equipment, *102–106*, 109; types of, *102–106*
median, **71**
meridians, **26**
microhabitat, **4–5**
Migratory Bird Conservation Act, 82
Migratory Bird Hunting and Conservation Stamp Act, 82
Migratory Bird Treaty Act, 82
mode of action, **136.** *See also* herbicide applications
modeling, 200–201
movement rate, 145

National Biology Manual (NBM), 184
National Environmental Policy Act (NEPA), 152–53, 173, 187
National Fire Danger Rating System (NFDRS), 89

National Forest Management Act (NFMA), 188
National Marine Fisheries Service (NMFS), 173–75
National Oceanic and Atmospheric Administration (NOAA), *170–71*
National Park Service (NPS), *170–71*, 187–88
National Park Service Organic Act, 187
National Planning Procedures Handbook (NPPH), 184–*85*
national wildlife refuge (NWR) system, 183
National Wildlife Refuge System Improvement Act, 183
Natural Resources Conservation Service (NRCS), 90, 108, 176, 184
naturalist, 1–2
Nature Conservancy, The (TNC), 167, *191*
nearest-neighbor method, 34
niche, **4–5**
Notice of Intent (NOI), 182–83

oaks (*Quercus*), 37
ocular estimates, 34–36, 38
optimum sustained yield, 130–31
outliers, **68**

pacing, **26**
parasites, 14, 126, 136, 205
pathogens, 14, 85, 113, 205
permeability, *12*
pesticides, 79–80, 173
photography, aerial, 22, 32, 56
Pittman-Robertson Act, 172, 178
plant succession, 5
plot, 33
plotless method, 33–34
plow as management tool. *See* herbicide applications
point counts, 33, 35, 49–50
point-intercept method, 35
point-quarter method, 34, 51–*52*
pollution: non-point source, **145**, 178, point, 178; water, 79, 82, 201–203
population: abundance, **22**, 39–41, 45–46, 48; accurate measurement, **23**; changes to wildlife, 9; closure, **22–23**, 45; density, **22**, 39–43, 45, 48–49; distribution, 43, 145; esti-

mate, **22**, 43, 46, 51, 53; frequency, 42–43; index, **23**, 40, 53; open, **23**, 46; precise measurement, **23**
poxvirus, 205
prairie potholes, 200
predators, 9, 14, 16, 135, 147
probability proportional to size (PPS) method, 51
protected species. *See* endangered species; threatened species
public participation. *See* stakeholder involvement

quadrant, 33–34
quadrat counts, 48

random pairs method, 34
rangelands, 76, 85–86, 107, 122–23, 130, 135–36, 142
Ray, John, 1
red-cockaded woodpecker (*Picoides borealis*), 118, 155, 207
relative humidity, **88**–89
remote cameras, 54–55
remote sensing, **31**–32, 40–41, 56–57, 194
Resource Management Plans (RMP), 187
Rio Grande silvery minnow (*Hybognathus amarus*), 206–207
Rivers and Harbors Act, 186
Rural Fire Prevention and Control, 182

salt cedar (*Tamarix* spp.), 32, 202–204
sampling design: accessibility (convenience sampling), 24; haphazard, 24; judgmental, 24; probabilistic, 43; simple random, 23–24, 43; stratified random, 24, 43; systematic (systematic-random), 24
Safe Harbor Agreements, 174–75
scale, **31**
scale, dimensions of, 6–7, 8
Schnabel estimator, 46–47
Schumacher-Eschmeyer estimator, 47
scrublands. *See* rangelands
sea-level rise, 44, *70*, 200–201
secretary of agriculture, 182, 188
seed germination, 85
significance, biologically, **64**
Sikes Act, 189